Coatings as Key Materials in Catalytic Applications

Coatings as Key Materials in Catalytic Applications

Guest Editor

Wanping Chen

MDPI

Basel • Beijing • Wuhan • Barcelona • Belgrade • Novi Sad • Cluj • Manchester

Guest Editor
Wanping Chen
School of Physics and Technology
Wuhan University
Wuhan
China

Editorial Office
MDPI AG
Grosspeteranlage 5
4052 Basel, Switzerland

This is a reprint of the Special Issue, published open access by the journal *Coatings* (ISSN 2079-6412), freely accessible at: https://www.mdpi.com/journal/coatings/special_issues/Catalytic_Coatings_Materials.

For citation purposes, cite each article independently as indicated on the article page online and as indicated below:

Lastname, A.A.; Lastname, B.B. Article Title. *Journal Name* **Year**, *Volume Number*, Page Range.

ISBN 978-3-7258-8261-8 (Hbk)
ISBN 978-3-7258-8262-5 (PDF)
https://doi.org/10.3390/books978-3-7258-8262-5

Contents

About the Editor vii

Michihisa Fukumoto, Hiroki Takahashi, Dawid Kutyła, Marek Wojnicki and Piotr Żabiński
Catalytic Activity Evaluation of Molten Salt-Treated Stainless Steel Electrodes for Hydrogen Evolution Reaction in Alkaline Medium
Reprinted from: *Coatings* **2024**, *14*, 796, https://doi.org/10.3390/coatings14070796 1

Senhua Ke, Chenyue Mao, Ruiqing Luo, Zeren Zhou, Yongming Hu, Wei Zhao and Wanping Chen
Surprising Effects of Al_2O_3 Coating on Tribocatalytic Degradation of Organic Dyes by CdS Nanoparticles
Reprinted from: *Coatings* **2024**, *14*, 1057, https://doi.org/10.3390/coatings14081057 12

Nengxun Yang, Yixuan Niu, Bohang Zhang and Fuchun Zhang
Photocatalytic Degradation of Sulfamethoxazole by Cd/Er-Doped Bi_2MoO_6
Reprinted from: *Coatings* **2024**, *14*, 1112, https://doi.org/10.3390/coatings14091112 23

Xiaoyu Yang, Ying Liu, Lin Wu, Zhiyuan Liao, Baoguo Zhang, Tinashe Tembo, et al.
Metal Ions' Dynamic Effect on Metal-Assisted Catalyzed Etching of Silicon in Acid Solution
Reprinted from: *Coatings* **2024**, *14*, 1405, https://doi.org/10.3390/coatings14111405 38

Guoqing Zhang, Jiankun Li, Yixing Wang, Linfeng Lei and Linzhou Zhuang
Controlled Aggregation of Cobalt and Platinum Atoms via Plasma Treatment for Exceptional Hydrogen Evolution Reaction Activity
Reprinted from: *Coatings* **2024**, *14*, 1569, https://doi.org/10.3390/coatings14121569 48

Fumei Lang, Xue Wen, Jibo Liu, Yineng Huang, Lili Zhang, Haiming Lu, et al.
First-Principles Study on the Electronic Structure and Optical Properties of $BiOIO_3$ Doped with As, Se, and Te
Reprinted from: *Coatings* **2025**, *15*, 111, https://doi.org/10.3390/coatings15010111 59

Xingping Fan, Ying Xia, Wei Fan and Yulong Li
Preparation and Photocatalytic Performance of Silver-Loaded Micro-Arc Oxidation TiO_2 Coating
Reprinted from: *Coatings* **2025**, *15*, 272, https://doi.org/10.3390/coatings15030272 72

Ruidan Duan, Jianhang Ding, Jiawei Fan and Linzhou Zhuang
Production of Self-Supporting Hollow Carbon Nanofiber Membranes with Co/Co_2P Heterojunctions via Continuous Coaxial Co-Spinning for Efficient Overall Water Splitting
Reprinted from: *Coatings* **2025**, *15*, 772, https://doi.org/10.3390/coatings15070772 86

About the Editor

Wanping Chen

Wanping Chen obtained his Bachelor's, Master's and Doctoral degrees from Tsinghua University, Huazhong University of Science and Technology, and Tsinghua University, respectively. From May 1998 to May 2001, he worked at the Institute for Materials Research, Tohoku University, Japan. Between May 2001 and December 2004, he was affiliated with the Department of Applied Physics, The Hong Kong Polytechnic University. He joined the School of Physics and Technology, Wuhan University, as a professor since 2005. For over 30 years, he has been dedicated to the synthesis and performance research of a variety of metal oxide functional materials. In 2018, his research interests began to cover the tribocatalytic performance of metal oxide semiconductors. His team was the first worldwide to report the tribocatalytic degradation of organic pollutants and the tribocatalytic reduction of carbon dioxide. In recent years, his team has made remarkable progress in establishing mechanisms for tribocatalysis, and they have also successfully upgraded relevant experimental facilities for tribocatalysis.

coatings

MDPI

Article

Catalytic Activity Evaluation of Molten Salt-Treated Stainless Steel Electrodes for Hydrogen Evolution Reaction in Alkaline Medium

Michihisa Fukumoto [1,*], Hiroki Takahashi [1], Dawid Kutyła [2,*], Marek Wojnicki [2] and Piotr Żabiński [2]

[1] Graduate School of Engineering Science, Akita University, Akita 010-8502, Japan; tkhshrk@gipc.akita-u.ac.jp
[2] Faculty of Non-Ferrous Metals, AGH University of Science and Technology, 30-059 Kraków, Poland; marekw@agh.edu.pl (M.W.); zabinski@agh.edu.pl (P.Ż.)
* Correspondence: fukumoto@gipc.akita-u.ac.jp (M.F.); kutyla@agh.edu.pl (D.K.)

Abstract: The goal of this research is to fabricate a novel type of highly active porous electrode material, based on stainless steel and dedicated to water electrolyzers. The main novelty of the presented work is the innovative application of the molten salts treatment, which allows the design of a highly developed porous structure, which characterizes significantly higher catalytic activity than untreated steel substrates. The equimolar mixture of NaCl and KCl with 3.5 mol% AlF_3 was used as the molten salt. The surface modification procedure includes the deposition of an Al layer with application at the potential of −1.8 V and following dissolution at −0.9 V, to create a porous alloy surface. The cathodic polarization measurements of the prepared porous stainless steel electrodes were measured in a 10 mass% KOH solution. Moreover, the amount of hydrogen generated during constant voltage electrolysis with a hydrogen sensor in situ was also measured. The porous stainless steel alloy showed higher current density at lower potentials in the cathodic polarization compared to untreated stainless steel. The cathodic polarization measurements in alkaline solution showed that the porous 304 stainless steel alloy is an excellent cathode material.

Keywords: molten salts modification; porous electrodes; stainless steel; hydrogen production; gas sensor

1. Introduction

The use of fossil fuels generates greenhouse gases, contributing to global warming, a major problem. Consequently, efforts are underway to mitigate greenhouse gas emissions. Hydrogen energy, which does not emit greenhouse gases, is gaining attention [1,2]. Alkaline water electrolysis, using renewable energy, is a method for producing this hydrogen, termed green hydrogen. However, production costs are high, necessitating cost-effective methods for alkaline water electrolysis. Therefore, there is a need to produce hydrogen efficiently and at a low cost.

Cathode electrodes have been developed to achieve high-efficiency hydrogen generation [3–20]. Historically, Ni alloys were predominantly used due to their corrosion resistance in alkaline water [8–19]. However, stainless steel has replaced Ni alloys [21–25]. The reason for this is that Ni is an expensive material, is provided a stable supply, and is difficult to process, so there is a need to develop inexpensive materials for electrodes used in alkaline water electrolysis. Therefore, stainless steel, which exhibits high corrosion resistance even in aqueous alkaline solutions, is being used. P.A. Selembo et al. are developing a low-cost cathode electrode made of stainless steel. The results show that stainless steel has higher hydrogen generation performance than Pt [21]. Y. Zhang et al. also investigated the hydrogen generation amount of stainless steel cathode electrodes and showed that stainless steel mesh gave good results [22]. Therefore, to enhance the properties of this stainless steel alloy, it is necessary to make it porous.

The authors have successfully generated porous surface layers of various alloys using an ionic liquid, molten salt [26,27]. In this study, Al was electrodeposited, as in previous studies by the authors, and aluminide was formed on the stainless steel surface. Therefore, we will investigate the hydrogen generation behavior of the porous stainless steel surface layer in an alkaline aqueous solution by dissolving Al after Al electrodeposition. However, in situ measurement of hydrogen generated by alkaline water electrolysis is challenging. The authors previously conducted in situ measurements of hydrogen generated by steam oxidation of metals using a proton conductor—a solid electrolyte—and the results showed accurate determination of trace amounts of hydrogen generated during the tests [28]. Moreover, this type of electrode can be successfully applied as a highly active material for oxidation of organic compounds [29]. In this study, we fabricated porous stainless steel using Al electrodeposition and Al dissolution in molten salt, aiming to evaluate its cathode performance. The surface and cross section of the fabricated samples were observed using SEM. To assess cathode performance, we measured the cathode polarization curve in 10 mass% KOH and the amount of hydrogen generated in situ using a hydrogen sensor fabricated with a solid electrolyte. We also elucidated the relationship between the amount of hydrogen generated and the porous stainless steel.

2. Experimental Section

SUS304 and SUS316 (The Nilaco Corporation, Tokyo, Japan) were utilized as base material samples. Table 1 displays the composition of each alloy. The sample surface was polished with emery paper up to #800 grit and subsequently cleaned using ultrasonic cleaning in acetone. The surface area of the sample is approximately 2 cm^2.

Table 1. Chemical composition of SUS304 and SUS316L (mass%).

	Cr	Ni	Mn	Si	Mo	Fe
SUS304	17%–19%	8%–11%	<2%	<1%		Balance
SUS316L	16%–18%	10%–14%	<2%	<1%	2%–3%	Balance

Figure 1 shows the formation procedure of porous stainless steel. After the deposition of aluminum (Al), a porous layer is created by selectively dissolving only Al. Aluminum electrodeposition cannot be conducted in an aqueous solution, therefore a molten salt was selected as the medium. For the electrolytic bath, an equimolar NaCl-KCl mixed salt containing 3.5 mol% AlF_3 was used. Following Al electrodeposition, only Al was dissolved in the same molten salt to generate a porous surface.

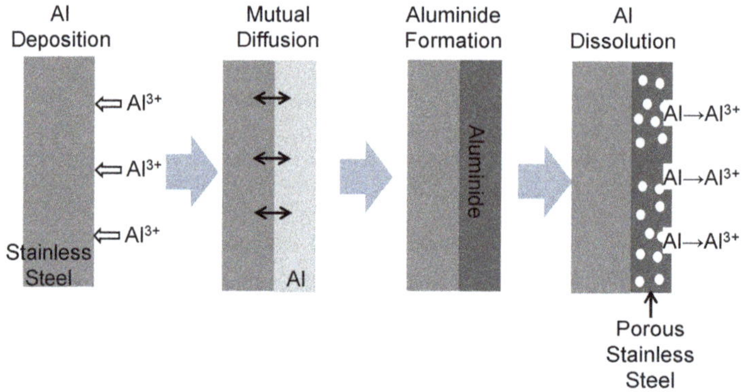

Figure 1. Schematic diagram of the formation procedure of porous stainless steel.

The electrolysis cell employed in this experiment is detailed in a previous report [24]. The reference electrode was a mullite tube with an outer diameter of 6 mm and a length of 500 mm, housing NaCl-KCl-AgCl (45:45:10 mol%) mixed salt with an immersed Ag wire. The melt temperature during Al electrodeposition and Al dissolution was maintained at 750 °C. Throughout the experiment, argon (Ar) gas was introduced into the cell at a flow rate of 200 cc min^{-1}.

Al deposition was executed at −1.8 V for 60 min using constant potential electrolysis. Subsequently, Al dissolution occurred at −0.9 V in all samples. The dissolution time continued until the current density reached zero. Post-treatment, the sample was removed from the bath, and salts that adhered to the sample surface were removed by washing with water.

The cross section of the treated sample underwent observation and analysis using a scanning electron microscope (SEM, Tokyo, Japan) and an X-ray micro-analyzer (Electron Probe Micro-Analyzer: EPMA, Tokyo, Japan). Furthermore, the deposits were identified using X-ray diffraction, with CuKα rays employed as the X-ray source.

To evaluate the hydrogen generation behavior of the sample after treatment, the cathodic polarization measurements were performed in a 10 mass% KOH solution. The potential sweep rate was 100 mV min^{-1}, and the temperature of the solution was maintained at 30 °C. The potential was swept in the cathode direction from the open circuit potential, which stabilized after immersion in solution, and measurements were taken until the potential reached −1.25 V.

Furthermore, the amount of hydrogen generated during electrolysis was measured in situ using a gas sensor. The measuring device and principle are detailed in a previous report [28]. Electrolysis was performed by applying the voltage equal 4.0 V in a 10 mass% KOH aqueous solution. Then, the amount of hydrogen generated at the cathode was measured by a carrier gas (Ar gas) and a gas sensor.

The amount of hydrogen generated was then calculated from the measured hydrogen partial pressure. If it is assumed that the measurement gas is an ideal gas and that Charles's law holds true, then Equation (1) holds true.

$$P_{\mathrm{H2}}(\mathrm{d}V\mathrm{r}/\mathrm{d}t) = (\mathrm{d}n/\mathrm{d}t)RT\mathrm{r} \quad (1)$$

where P_{H2} is the measured hydrogen pressure, $V\mathrm{r}$ is the gas flow rate measured at the temperature $T\mathrm{r}$, n is the amount of gas generated, and $T\mathrm{r}$ is the temperature at which the flow rate was measured. In this study, the temperature was 25 °C. The amount of hydrogen generated per unit time was calculated, and the total amount of hydrogen generated was calculated at each temperature.

3. Results and Discussion

3.1. Anodic Polarization Curves of Various Metals

Figure 2 illustrates the anodic polarization curves of various metals, measured in NaCl-KCl-3.5 mol% AlF_3 molten salt at 750 °C. Concerning aluminum (Al), experiments were conducted using samples where Al was electrodeposited on stainless steel. The obtained data reveal that Al dissolves at the lowest potential, while Ni dissolves at the highest potential. To induce the dissolution reaction of Al selectively, constant polarization at −0.9 V seems to be effective. In the case of chromium (Cr), an increase in anodic current, indicating the dissolution process, was observed from around −0.85 V. Additionally, for iron (Fe), an increase in anodic current was noted from around −0.75 V. These findings underscore that Al exhibits the highest solubility, whereas Ni exhibits the highest resistance for electrochemical dissolution in molten salts mixture.

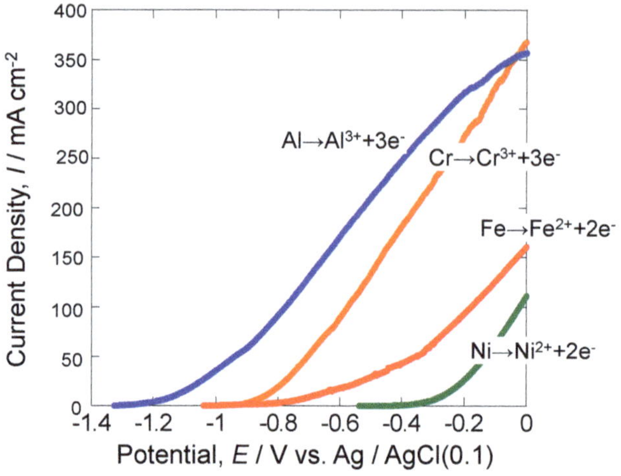

Figure 2. Anodic polarization curves of various metals in the 50 mol%NaCl-50 mol%KCl molten salt at 750 °C.

3.2. Current Density–Time Curve during Al Deposition and Al Dissolution onto SUS304 and SUS316L Stainless Steel Substrates

Figure 3 shows the current density–time curves during both Al electrodeposition and Al dissolution. Al was deposited at −1.8 V for 1 h in NaCl-KCl-3.5 mol% AlF_3 molten salt at 750 °C. Subsequently, only Al was dissolved at −0.9 V. The figures illustrate that a cathodic current flows during electrodeposition, while an anodic current flows during dissolution. Notably, the experiment was sustained until the dissolution current reached zero, indicating the complete dissolution of all Al. During electrodeposition, an initially high cathode current is evident, followed by a stabilization around −120 mA cm^{-2} at 60 min for SUS304 and around −150 mA cm^{-2} at 60 min for 316L. The dissolution current exhibited an initial peak, followed by a rapid decline. Subsequently, the dissolution reaction of Al occurred at a consistent rate until the current value reached zero.

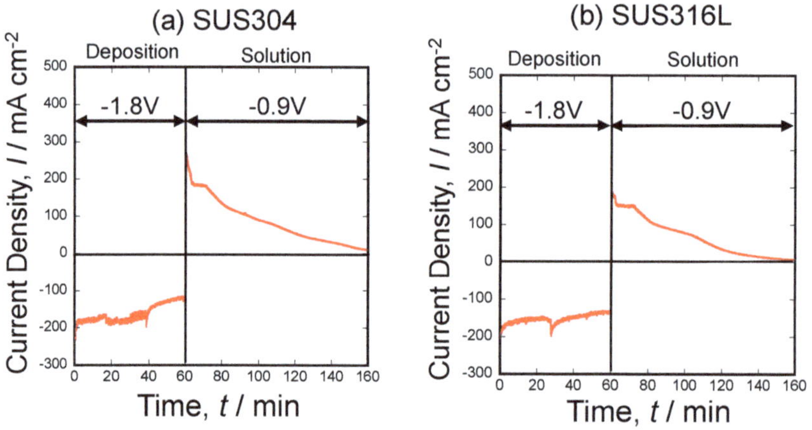

Figure 3. Time dependence of current by the constant potential polarization at 750 °C in the 48.25 mol%NaCl-48.25 mol%KCl-3.5 mol%AlF_3 molten salt. (**a**) SUS304; (**b**) SUS316L.

3.3. Morphology of Stainless Steel Samples after Al Molten Salts Electrodeposition

Figure 4 displays scanning electron microscope (SEM) images revealing the surface characteristics of the samples subsequent to the electrodeposition of aluminum (Al) at 750 °C and −1.8 V in NaCl-KCl-3.5 mol% AlF_3 molten salt. Two types of stainless steel, SUS304 and SUS316L, were employed in this process, and the outcomes are detailed below.

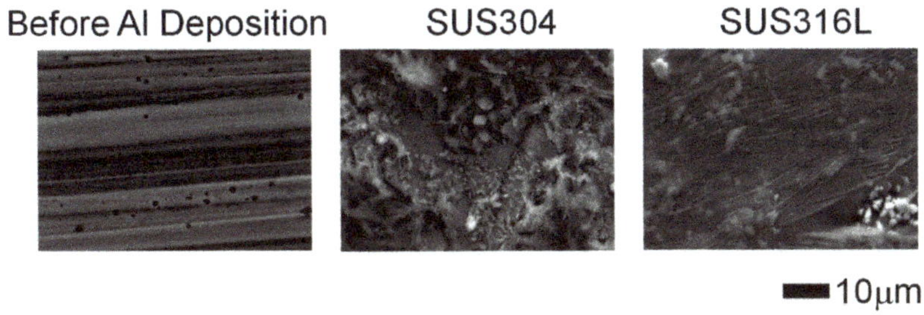

Figure 4. Morphology of sample surface after Al electrodeposition with constant potential polarization at −1.8 V at 750 °C for 1 h.

Following the Al electrodeposition, the samples were cleaned via ultrasonic cleaning in water to eliminate any molten salt residues adhering to the surface. The samples were dried and then used for observation. Numerous irregularities were observed on the surfaces of both SUS304 and SUS316L samples. Importantly, no defects such as voids or empty spaces were detected. Moreover, it is evident that the surface has acquired a solidified structure due to the electrodeposition of liquid Al. This observation confirms that no voids or empty spaces were present on the sample surface after the Al electrodeposition process.

Figure 5 presents scanning electron microscope (SEM) images captured using backscattered electron mode, shows the cross-sectional views of two stainless steel samples after the deposition of aluminum (Al) at −1.8 V in NaCl-KCl-3.5 mol% AlF_3 molten salt at 750 °C. In both samples, an Al-rich Fe aluminide layer was observed on the surface. This occurrence is related to the electrodeposited Al being in a liquid state and diffusing into the substrate. Furthermore, a substantial deposited layer was formed in the 316L sample. This layer had an approximate thickness of 60 μm, exhibited good adhesion to the stainless steel, and presented a uniform deposition across the sample surface. The cross-sectional views provide valuable insights into the structural changes induced by the Al deposition process in the stainless steel samples.

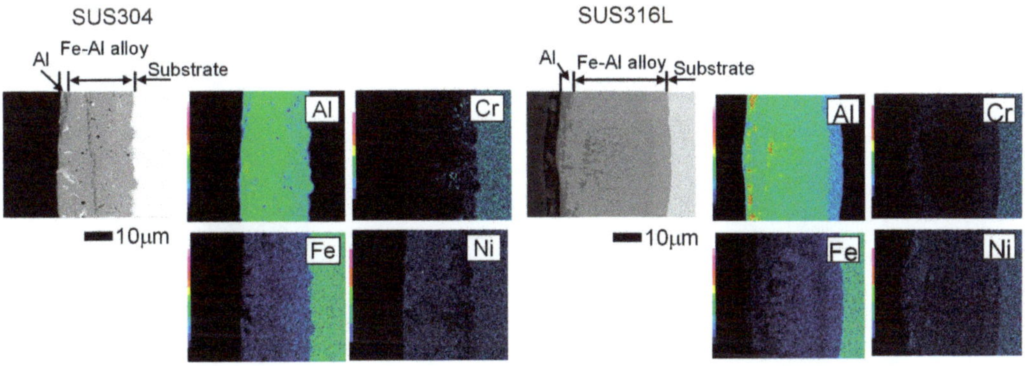

Figure 5. Cross section of microstructure after Al electrodeposition with constant potential polarization at −1.8 V at 750 °C for 1 h.

3.4. Morphology and Composition of Porous Steel Samples after Molten Salts Treatment and Dissolution of Al

Figure 6 illustrates the surface morphology of two different steel samples subjected to molten salts treatment by the electrodeposition and dissolution of aluminum. The Al deposition conditions were set at −1.8 V for 60 min, and the Al dissolution conditions were maintained at −0.9 V for both samples. On the surface, many voids were observed with irregular shape. Finer voids were registered for SUS304, while in 316L, the voids exhibited a more complex structure. The porous morphology is strongly dependent on the type of stainless steel used for molten salt treatment.

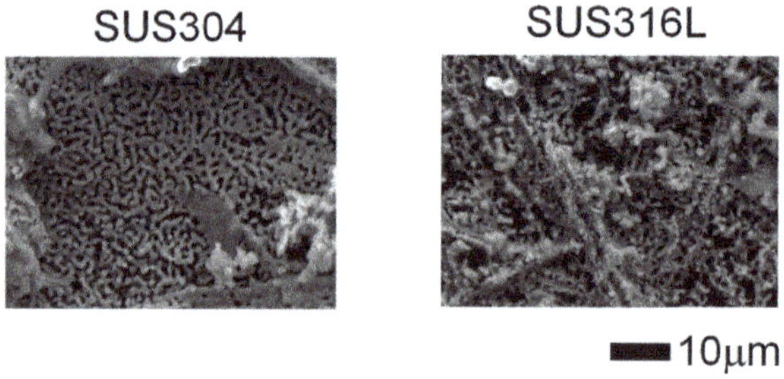

Figure 6. Morphology of sample surface after Al electrodeposition and dissolution with constant potential polarization at −1.8 V to −0.9 V at 750 °C.

In Figure 7, the cross-sectional structure of the two porous samples after treatment by the electrodeposition and dissolution of Al is presented. A fine porous layer is evident in both samples, with a thickness of approximately 50 mm in each case. Analysis revealed no presence of Al in the porous layer, indicating complete dissolution and removal of the electrodeposited Al. The obtained microstructure of the steel-based samples seems to be more suitable and compact than in other types of molten salt-modified materials obtained in our previous studies: Co-Ni, Ni, and Ni-Pt. This outcome can be related with the presence of different microstructural carbides and intermetallics, typical for steel, which are highly resistant to phase formation with molten Al and the further dealloying process.

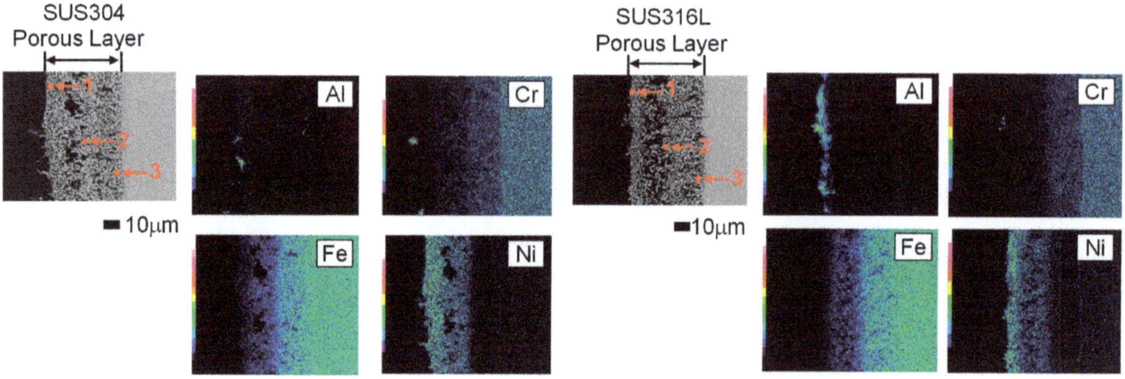

Figure 7. Cross-sectional microstructure of sample after Al electrodeposition and dissolution with constant potential polarization at −1.8 V to −0.9 V at 750 °C. 1 Top Part, 2 Middle Part, 3 Under Part.

Table 2 displays point analysis results of the cross section shown in Figure 7. In SUS304 steel, at point 1, Cr concentration was 6.81 at.%, Fe was 21.9 at.%, and Ni was 71.29 at.%, indicating a concentration of Ni on the surface. Moreover, point 2 on the substrate side, Cr concentration was 13.2 at.%, Fe was 55.4 at.%, and Ni was 26.4 at.%, with the Ni concentration decreasing rapidly. This observation suggests an increase in Ni concentration on the surface of SUS304 with porous treatment. In case of SUS316L, the point 1, Cr was 1.78 at.%, Fe was 11.09 at.%, and Ni was 87.14 at.%, revealing a higher Ni concentration than in SUS304. At point 2 on the substrate side, Cr was 10.86 at.%, Fe was 53.51 at.%, and Ni was 35.63 at.%, with a rapid decrease in Ni concentration. The surface Ni concentration varied depending on the type of stainless steel, potentially influencing the behavior of hydrogen generation.

Table 2. Point analysis results at each point in Figure 7.

	SUS304 (at.%)				SUS316L (at.%)		
	Cr	Fe	Ni		Cr	Fe	Ni
1	6.81	21.90	71.29	1	1.78	11.09	87.14
2	13.2	55.4	26.4	2	10.86	53.51	35.63
3	13.91	58.38	27.70	3	14.96	72.31	12.73

Figure 8 presents the XRD diffraction patterns obtained for samples after Al electrodeposition and dissolution of the porous-treated sample. The formation of the Fe-Al phase occurred during Al deposition at −1.8 V and 750 °C for 60 min in both samples. The subsequent dissolution of Al and the porous treatment resulted in a profile similar to that of the substrate material, indicating that the formation of the porosity in the material is not changing the original crystalline structure even after treatment in molten salt at 750 °C.

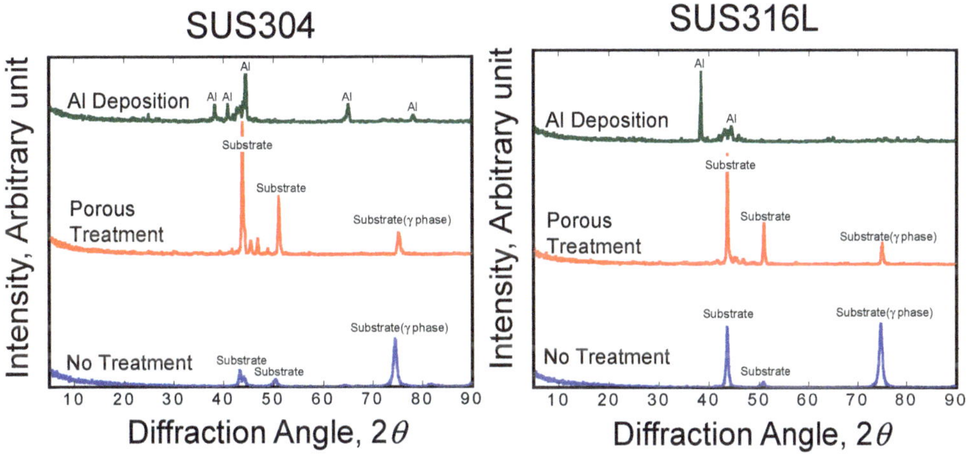

Figure 8. X-ray diffraction patterns registered for substrate, Al-deposited, and porous samples obtained for SUS304 (**left**) and SUS316L (**right**) stainless steel.

3.5. Electrochemical Activity Evaluation of Porous Stainless Steel Electrodes Obtained by Molten Salts Treatment in Hydrogen Evolution Reaction

Figure 9 presents the results of cathodic polarization scans measured in a 10 mass% KOH solution for both samples: untreated substrate and stainless steel after molten salts treatment. It should be noted that the cathodic current density registered in polarization scans differs significantly between the porous samples and the non-treated substrate. In untreated SUS304, an increase in cathodic current was observed when the voltage was below −1.4 V. Conversely, with porous treatment, a rise in cathodic current was noted from −1.1 V. Similarly, in untreated SUS316L, an increase in cathode current occurred from around −1.35 V, while porous treatment led to an increase from around −1.1 V. Both porous-treated samples reach a cathode current of 60 mA cm^{-2} at −1.4 V for SUS304 and −1.44 V for SUS316L, indicating that porous SUS304 achieves this current at −1.4 V. This highlights a substantial change in the catalytic activity for hydrogen evolution, which can be observed under more positive potential values than in the case of non-treated steel substrates.

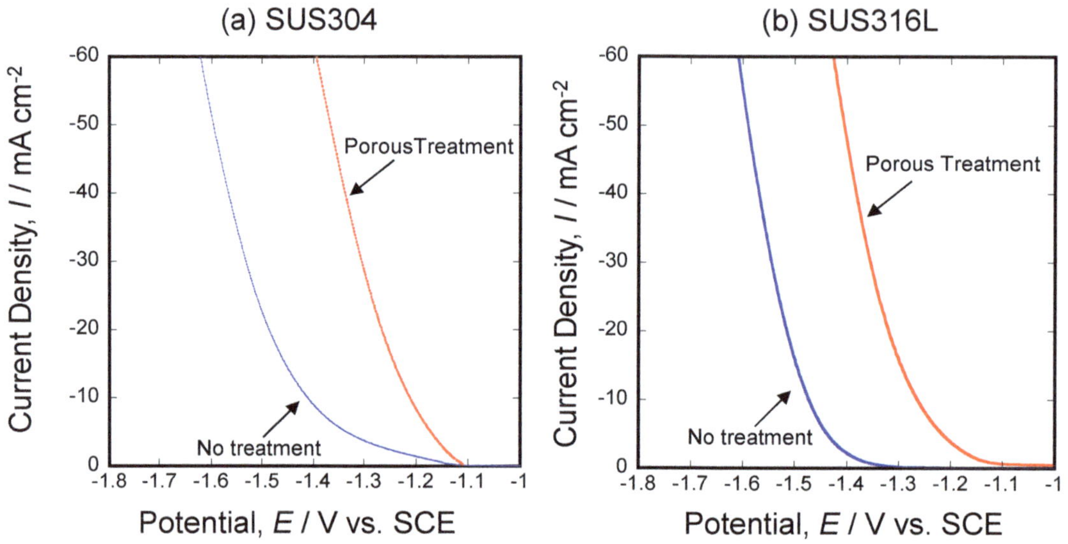

Figure 9. Cathodic polarization scans registered for different types of steel substrates and porous samples in 10 mass% KOH solution. (**a**) SUS304; (**b**) SUS316L.

Figure 10 presents the hydrogen partial pressure (Figure 10a) measured with a hydrogen sensor and the hydrogen partial pressure calculated using the ideal gas equation for hydrogen generated at the cathode electrode during electrolysis at 4.0 V in 10 mass% KOH, showing the amount of hydrogen generated (Figure 10b). Equation (1) was used to calculate the amount of hydrogen generated from the hydrogen partial pressure, as shown in the experimental method section. In Figure 10a, the untreated sample exhibited a low hydrogen partial pressure, but this pressure increased significantly with porous treatment. Particularly, a rapid rise in hydrogen partial pressure was observed initially in SUS304 steel, indicating that porous treatment enhances the amount of hydrogen generated during electrolysis. It was also evident that hydrogen was rapidly generated in the initial stage of treated SUS304 steel. However, the temporal change in the amount of hydrogen generated, calculated from the hydrogen partial pressure, revealed that the amount of hydrogen generated was lower for both samples than in the untreated sample. Nevertheless, it is evident that the amount of hydrogen generated increases significantly through treatment. By integrating Figure 10b, the total amount of hydrogen generated can be calculated. As a result, it was revealed that the untreated SUS304, which showed the lowest value, had a value of 1.8×10^{-11} mL, but with treatment, it reached 4.0×10^{-11} mL, more than double the initial amount. Comparing the overvoltage (10 mA cm^{-2}) in past papers, the sample created in this research has an overvoltage of 80 mV, which is considered a good result [30]. These findings highlight the successful fabrication of a porous stainless steel surface and the consequential improvements in cathode performance and hydrogen generation efficiency. The observed characteristics indicate the potential applications of porous stainless steel in efficient hydrogen production processes.

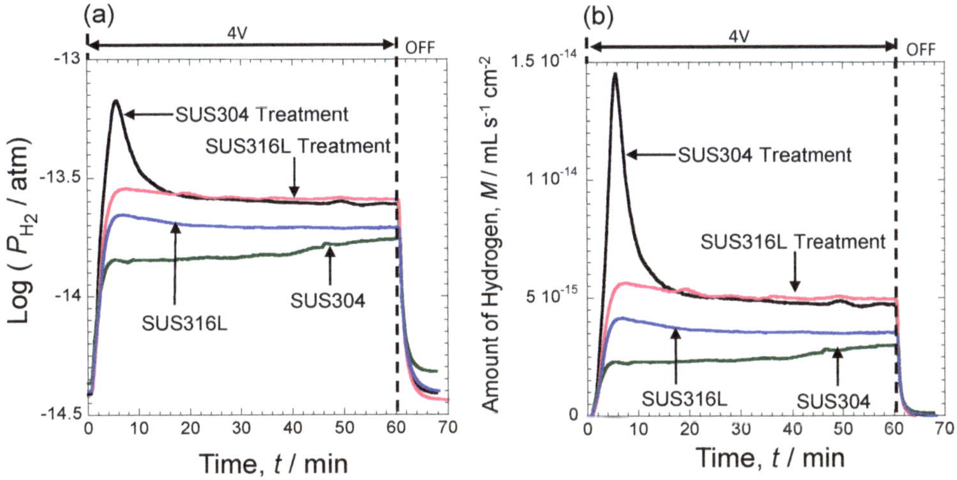

Figure 10. Time dependence of hydrogen partial pressure (**a**) and amount of hydrogen generated (**b**) during electrolysis at 4 V.

4. Conclusions

Two types of stainless steel were employed as substrates where aluminum was deposited from a molten salt electrolyte to create a layer of Fe-Al alloy. Subsequently, the deposited Al was liquefied under high-temperature conditions, resulting in the penetration into the stainless steel substrate. The solid-state reaction between Al and Fe led to formation of a mixture of intermetallic phases. Obtained Al-rich phases were electrochemically dissolved with the application of −0.9 V potential, revealing a porous stainless steel alloy layer.

Furthermore, the cathodic performance of the prepared porous stainless steel alloy surface layer was tested and compared with untreated stainless steel substrates. Additionally, the amount of hydrogen generated during constant voltage electrolysis in a 10 mass% KOH solution was quantified using a hydrogen gas sensor. The obtained results of the conducted study can be summarized as follows:

- A porous stainless steel surface was successfully fabricated through Al electrodeposition and Al dissolution experiments.
- It was observed that porous treatment allows to obtain a compact surface with numerous voids.
- The porous stainless steel alloy exhibited a higher current density at lower potentials in the cathodic polarization curve compared to the untreated sample
- In a hydrogen generation experiment using constant voltage electrolysis, the steel samples after porous treatment produced more hydrogen, especially SUS304 which generated the highest amount of hydrogen and produced 4×10^{-11} mL of hydrogen on a surface area of 2 cm^2 in 1 h of electrolysis time.

Author Contributions: Conceptualization, M.F., H.T., D.K., M.W. and P.Ż.; methodology, M.F.; formal analysis, M.F., H.T. and D.K.; investigation, M.F., H.T., D.K. and M.W.; data curation, M.F., H.T. and D.K.; writing—original draft preparation, M.F.; writing—review and editing, H.T., D.K., M.W. and P.Ż.; supervision, M.F.; project administration, M.F., H.T., D.K., M.W. and P.Ż. All authors have read and agreed to the published version of the manuscript.

Funding: This research received no external funding.

Institutional Review Board Statement: Not applicable.

Informed Consent Statement: Not applicable.

Data Availability Statement: The original contributions presented in the study are included in the article, further inquiries can be directed to the corresponding authors.

Conflicts of Interest: The authors declare no conflict of interest.

References

1. Momirlan, M.; Veziroglu, T.N. The properties of hydrogen as fuel tomorrow in sustainable energy system for a cleaner planet. *Int. J. Hydrogen Energy* **2005**, *30*, 795–802. [CrossRef]
2. Safizadeh, F.; Ghali, E.; Houlachi, G. Electrocatalysis developments for hydrogen evolutuion reaction in alkaline solutions—A Review. *Int. J. Hydrogen Energy* **2015**, *40*, 256–274. [CrossRef]
3. Chen, L.; Lasia, A. Study of the kinetics of hydrogen evolution reaction on nickel-zinc alloy electrodes. *J. Electrochem. Soc.* **1991**, *138*, 3321–3328. [CrossRef]
4. Chen, L.; Lasia, A. Study of the kinetics of hydrogen evolution reaction on nickel-zinc powder electrodes. *J. Electrochem. Soc.* **1992**, *139*, 3214–3219. [CrossRef]
5. Rami, A.; Lasia, A. Kinetics of hydrogen evolution on Ni-Al alloy electrodes. *J. Appl. Electrochem.* **1992**, *22*, 376–382. [CrossRef]
6. Birry, L.; Lasia, A. Studies of the hydrogen evolution reaction on Raney nickel-molybdenum electrodes. *J. Appl. Electrochem.* **2004**, *34*, 735–749. [CrossRef]
7. Navarro-Flores, E.; Chong, Z.; Omanovic, S. Characterization of Ni, NiMo, NiW and NiFe electroactive coatings as electrocatalysts for hydrogen evolution in an acidic medium. *J. Mol. Catal. A Chem.* **2005**, *226*, 179–197. [CrossRef]
8. Mauer, A.E.; Kirk, D.W.; Thorpe, S.J. The role of iron in the prevention of nickel electrode deactivation in alkaline electrolysis. *Electrochim. Acta* **2007**, *52*, 3505–3509. [CrossRef]
9. Dong, H.; Lei, T.; He, Y.; Xu, N.; Huang, B.; Liu, C.T. Electrochemical performance of porous Ni_3Al electrodes for hydrogen evolution reaction. *Int. J. Hydrogen Energy* **2011**, *36*, 12112–12120. [CrossRef]
10. Mullis, A.M.; Bigg, T.D.; Adkins, N.J. A microstructural investigation of gas atomized Raney type Al-27.5 at.% Ni catalyst precursor alloys. *J. Alloys Compd.* **2015**, *648*, 139–148. [CrossRef]
11. Yu, L.; Lei, T.; Nao, B.; Jiang, Y.; He, Y.; Liu, C.T. Characteristics of a sintered porous Ni-Cu alloy cathode for hydrogen production in a potassium hydroxide solution. *Energy* **2016**, *97*, 498–505. [CrossRef]
12. Brown, I.J.; Sotiropoulos, S. Preparation and characterization of microporous Ni coatings as hydrogen evolving cathodes. *J. Appl. Electrochem.* **2000**, *30*, 107–111. [CrossRef]
13. Tanaka, S.; Hirose, N.; Tanaki, T.; Ogata, Y.H. The effect of tin ingredients on electrocatalytic activity of Raney-Ni prepared by mechanical alloying. *Int. J. Hydrogen Energy* **2001**, *26*, 47–53. [CrossRef]
14. Yüce, A.O.; Döner, A.; Kardaş, G. NiMn composite electrodes as cathode material for hydrogen evolution reaction in alkaline solution. *Int. J. Hydrogen Energy* **2013**, *38*, 4466–4473. [CrossRef]
15. Chade, D.; Berlouis, L.; Infield, D.; Cruden, A.; Nielsen, P.N.; Mathiesen, T. Evaluation of Raney nickel electrodes prepared by atmos pheric plasma spraying for alkaline water electrolysers. *Int. J. Hydrogen Energy* **2013**, *38*, 14380–14390. [CrossRef]
16. Lupi, C.; Dell'Era, A.; Pasquali, M. Nickel-cobalt electrodeposited alloys for hydrogen evolution in alkaline media. *Int. J. Hydrogen Energy* **2009**, *34*, 2101–2106. [CrossRef]
17. González-Buch, C.; Herraiz-Cardona, I.; Ortega, E.; García-Antón, J.; Pérez-Herranz, V. Synthesis and characterization of macroporous Ni, Co and Ni-Co electrocatalytic deposits for hydrogen evolution reaction in alkaline media. *Int. J. Hydrogen Energy* **2013**, *38*, 10157–10169. [CrossRef]
18. Wang, C.; Zhang, Q.; Liu, Z.; Li BZhao, W.; Zhang, C.; Jiang, S.; Wang, J.; Liu, K.; He, S. CoO supported NiFe layered double hydroxide sandwich-like nanosheets on hierarchical carbon framework for efficient electrocatalytic oxygen evolution. *ChemSusChem* **2024**, *17*, e202301703. [CrossRef] [PubMed]
19. Zhou, Z.; Springer, M.A.; Geng, W.; Zhu, X.; Li, T.; Li, M.; Jing, Y.; Heine, T. Rational Design of Two-Dimensional Binary Polymers from Heterotriangulenes for Photocatalytic Water Splitting. *J. Phys. Chem. Lett.* **2021**, *33*, 8134–8140. [CrossRef]
20. Dong, B.; Yu, N.; Wang, Q.Y.; Ren, J.K.; Zhang, X.Y.; Zhang, Z.J.; Fan, R.Y.; Liu, D.P.; Chai, Y.M. Double active sites promoting hydrogen evolution activity and stability of $CoRuOH/Co_2P$ by rapid hydrolysis. *Chin. Chem. Lett.* **2024**, *35*, 109221. [CrossRef]
21. Lavorante, M.J.; Franco, J.I. Performance of stainless steel 316L electrode with modified surface to be use in alkaline water electrolyers. *Int. J. Hydrogen Energy* **2016**, *41*, 9731–9737. [CrossRef]
22. David, M.; Ocampo-Martínez, C.; Sánchez-Pena, R. Advance in alkaline water electrolyzers: A review. *J. Energy Storage* **2019**, *23*, 392–403. [CrossRef]
23. Moureaux, F.; Stevens, P.; Toussaint, G. Timely-activated 316L stainless steel: A low cost, durable and active electrode for oxygen evolution reaction in concentrated alkaline environments. *Appl. Catal. B Environ.* **2019**, *258*, 117963. [CrossRef]
24. Zamanizadeh, H.R.; Sunde, S.; Pollet, B.G.; Seland, F. Tailoring the oxide surface composition of stainless steel for improved OER performance in alkaline water electrolysis. *Electrochim. Acta* **2022**, *424*, 140561. [CrossRef]
25. Zamanizadeh, H.R.; Barnett, A.O.; Sunde, S.; Pollet, B.G.; Seland, F. Performance of activated stainless steel and nickel-based anodes in alkaline water electrolyser. *J. Power Sources* **2023**, *564*, 232828. [CrossRef]
26. Fukumoto, M.; Sugiuchi, K.; Nakajima, K. Formation of porous Ni surface by electrodeposition and dissolution in molten salt. *Int. J. Hydrogen Energy* **2020**, *45*, 28252–28259. [CrossRef]

27. Nakajima, K.; Fukumoto, M. Porous Ni-Co surface formation and analysis of hydrogen generation by gas sensor. *Int. J. Hydrogen Energy* **2021**, *46*, 26263–26271. [CrossRef]
28. Fukumoto, M.; Nakajima, K.; Kawamori, Y. Investigation of Alumina Formation and Oxidation Rate of Ni-5wt%Al-Xwt%Cr Alloy Using Hydrogen Sensor and Oxygen Pump Sensor. *Oxid. Met.* **2020**, *94*, 191–204. [CrossRef]
29. Fukumoto, M.; Takahashi, H.; Kutyla, D.; Wojnicki, M.; Zabinski, P. Morphological investigation and electrochemical performance evaluation of novel porous Ni-Pt produced by Al-deposition/dissolution in molten salts for hydrogen and oxygen evolution reaction. *Int. J. Hydrogen Energy* **2024**, *49*, 754–765. [CrossRef]
30. Huang, L.; Hou, Y.; Yu, Z.; Peng, Z.; Wang, L.; Huang, J.; Zhang, B.; Qian, L.; Wu, L.; Li, Z. Pt/Fe-NF electrode with high double-layer capacitance for efficient hydrogen evolution reaction in alkaline media. *Int. J. Hydrogen Energy* **2017**, *42*, 9458–9466. [CrossRef]

coatings

MDPI

Article

Surprising Effects of Al_2O_3 Coating on Tribocatalytic Degradation of Organic Dyes by CdS Nanoparticles

Senhua Ke [1], Chenyue Mao [1], Ruiqing Luo [1], Zeren Zhou [1], Yongming Hu [2], Wei Zhao [3] and Wanping Chen [1,*]

1 Key Laboratory of Artificial Micro- and Nano-Structures of Ministry of Education, School of Physics and Technology, Wuhan University, Wuhan 430072, China
2 Hubei Key Laboratory of Micro–Nanoelectronic Materials and Devices, School of Microelectronics, Hubei University, Wuhan 430062, China
3 School of Materials Science and Engineering, Tianjin Chengjian University, Tianjin 300384, China
* Correspondence: wpchen@whu.edu.cn

Abstract: With a band gap of 2.4 eV, CdS has been extensively explored for photocatalytic applications under visible light irradiation. In this study, CdS nanoparticles have been investigated for the tribocatalytic degradation of concentrated Rhodamine B (RhB) and methyl orange (MO) solutions. For CdS nanoparticles in a glass beaker, 78.9% of 50 mg/L RhB and 69.8% of 20 mg/L MO solutions were degraded after 8 h and 24 h of magnetic stirring using Teflon magnetic rotary disks, respectively. While for CdS nanoparticles in a beaker with Al_2O_3 coated on its bottom, 99.8% of the RhB solution was degraded after 8 h of magnetic stirring and 95.6% of the MO solution was degraded after 12 h of magnetic stirring. Moreover, another contrast was observed between the two beaker bottoms—a new peak at 250 nm in UV–visible absorption spectra was only observed for the MO degradation by CdS in the as-received glass beaker, which indicates that MO molecules were only broken into smaller organic molecules in that case. These findings are meaningful for expanding the catalytic applications of CdS and for achieving a better understanding of tribocatalysis as well.

Keywords: tribocatalysis; dye degradation; CdS; coating; Al_2O_3

1. Introduction

With the development of science and technology, the discharge of pollutants from all walks of life is constantly damaging the water resources on the earth [1]. The composition of polluted water is very complicated, often containing plastics, antibiotics, organic dyes, etc. [2–6]. Organic dyes not only destroy the ecosystem, but also are toxic to humans, leading to skin diseases, such as dermatitis and psoriasis, and even inducing malignant lesions, causing great negative effects [7]. Therefore, it is urgent to find simple, effective, and low-cost methods to treat polluted water and improve the living environment.

There exists a huge amount of clean energy in nature in the forms of solar energy [8], wind energy [9], geothermal energy [10], ocean energy [11], and so on. Presently, these forms of clean energy are mainly obtained from the environment and are converted into chemical energy and electricity [12]. Photocatalysis is a mainstream method for water pollution control by using solar energy as clean energy [13]. Electrons and holes are generated in photocatalysts by light irradiation, and the electrons and holes further participate in the REDOX reaction to decompose harmful substances in wastewater [14]. Although photocatalysis is a mature water remediation method, which has undergone several decades of development, it still faces some challenges and obstacles in practical application, including a high photogenic carrier recombination rate and a low visible light utilization rate [15–17]. Obviously, other forms of clean energy should be harnessed more to fill the gap of photocatalytic reactions in environmental remediation.

As mechanical energy is abundant and widely available in the ambient environment, it has received more and more attention to be collected through some catalytic technologies in recent years. In this context, tribocatalysis has emerged as an appealing technology in environmental remediation in recent years. In fact, the terminology of tribocatalysis was proposed decades ago. Heinicke et al. first defined tribocatalysis as a branch of tribochemistry, whose subject is the change in catalytic properties of solids under the action of mechanical energy [18]. Since then, tribocatalysis has been mostly studied for the promotion of tribochemical reactions, reducing friction, and achieving super lubrication [19–21]. In 2019, Li et al. first reported the tribocatalytic degradation of organic pollutant dyes by $Ba_{0.75}Sr_{0.25}TiO_3$ nanoparticles [22], in which $Ba_{0.75}Sr_{0.25}TiO_3$ nanoparticles degraded organic pollutants into pollution-free small molecules by collecting mechanical energy under the condition of magnetic stirring. Very quickly, tribocatalysis has also been reported for some materials to absorb mechanical energy via friction for the conversion of H_2O and CO_2 into flammable gasses [23,24]. Obviously, the scope of tribocatalysis has been extended from tribochemical reactions to the collection and conversion of mechanical energy.

Up to now, many materials investigated in tribocatalytic environmental remediation possess a semiconductor-type band gap, such as $CaCu_3Ti_4O_{12}$ (CCTO) [25], $BaTiO_3$ [26], TiO_2 [27], Si [28], Bi_2WO_6 [29], ZnO [30], $SrTiO_3$ [31], and $CoFe_2O_4$ [32]. Based on the excitation of electron-hole pairs in semiconductors by mechanical energy absorbed through friction, a mechanism has been established for tribocatalysis [23,24]. This mechanism is not only very similar to that of photocatalysis, but also suggests that those materials with outstanding photocatalytic properties may also be promising for tribocatalytic environmental remediation. It is well known that CdS is an important semiconductor material with a band gap of 2.4 eV, which is resistant to optical and chemical corrosion, absorbs a wide range of electromagnetic waves, and has been widely studied as a visible light photocatalyst to convert toxic chemicals into nontoxic small molecules through photocatalysis [33–35]. As a matter of fact, CdS is also among the earliest semiconductors that were investigated for tribocatalytic environmental remediation [36]. Nevertheless, CdS has only been investigated for the tribocatalytic degradation of rhodamine B (RhB) of a low concentration (5 mg/L) up to now; this is much easier to be degraded than many dye solutions that have appeared in tribocatalytic investigations. In this study, we have further explored the potential of CdS nanoparticles in the tribocatalytic degradation of some much more stubborn organic pollutants, including 50 mg/L RhB and 20 mg/L methyl orange (MO) solutions. Though it is quite challenging for CdS nanoparticles to degrade them through magnetic stirring in a normal way, degradation is found to be surprisingly enhanced through an Al_2O_3 coating on the beaker bottoms. Especially for the 20 mg/L MO solution, an Al_2O_3 coating not only dramatically increases the degradation speed, but also changes the degradation mode from a partial degradation to a full one. These results are important not only for tribocatalytic environmental remediation, but also for achieving a better understanding of tribocatalysis as a whole.

2. Materials and Methods

2.1. Materials and Characterization

Commercial CdS nanoparticles with a nominal purity of 99.99 wt% were used in this study, which were purchased from Shanghai Aladdin Biochemical Technology Co., Ltd. (Shanghai, China). The crystal structure of the CdS powder was measured through an X-ray diffractometer (XRD, SmartLab SE, Rigaku, Tokyo, Japan) using Cu Kα radiation. The morphology of the CdS powder was observed through a scanning electron microscope (SEM, Zeiss GeminiSEM 500, Oberkochen, Germany), and the microstructure was analyzed. X-ray photoelectron spectroscopy (XPS, Thermo escalab 250XI, Waltham Massachusetts, USA) was used to analyze the elemental composition and chemical state of CdS nanoparticles, and electron binding energies were calibrated using the reference peak of C 1s (284.6 eV).

2.2. Coating Al_2O_3 Ceramic Disks on the Bottoms of Glass Beakers

Commercial flat-bottomed glass beakers, φ 45 mm $\times$ 60 mm, were divided into two groups in this study. In the first group, the glass beakers were used directly; in the other group, Al_2O_3 ceramic disks of φ 40 mm × 1 mm were pasted on their bottoms using a strong glue (deli super glue 502) before they were used. In this way, we had two kinds of glass beakers with glass and Al_2O_3 bottoms, respectively.

2.3. Tribocatalytic Degradation of RhB and MO Solutions

For solutions of organic dyes, the higher the concentration, the more difficult it is to be degraded. In this study, relatively concentrated RhB (50 mg/L) and MO (20 mg/L) solutions were adopted to increase their degradation difficulty. In a typical experiment, 0.30 g of CdS nanoparticles were dispersed in a glass beaker containing 30 mL of either 50 mg/L RhB or 20 mg/L MO solution. A homemade Teflon magnetic rotary disk, which was described in detail in a previous paper [7], was used to magnetically stir the suspension at 400 rpm. The room temperature was kept at 25 °C, and the beaker was kept in the dark. During the test, 1 mL of the solution was taken at regular intervals and was centrifuged at 8000 rpm for 5 min to remove CdS nanoparticles. The absorbance of the solutions was measured through a UV–visible spectrometer (UV-2550; Shimadzu, Kyoto, Japan) over the range of 200–650 nm.

2.4. Detection of Radical Species

For hydroxyl radical detection, 10 mL of deionized water, 50 µL of 5, 5-dimethyl-1-pyrrolin-n-oxide (DMPO), and 0.15 g of CdS nanoparticles were added to two glass beakers (φ 45 mm × 60 mm) with either a glass or Al_2O_3 bottom, respectively. For the detection of superoxide radicals, 10 mL of methanol, 50 µL of 5, 5-dimethyl-1-pyrrolin-n-oxide (DMPO), and 0.15 g of CdS nanoparticles were added to two glass beakers with either a glass or Al_2O_3 bottom, respectively. A Teflon magnetic rotary disk was used in every beaker to stir the suspension in it at 400 rpm for 15 min in the dark at room temperature. An electron paramagnetic resonance (EPR) spectrometer (A300-10/12, Bruker, Berlin, Germany) was used to separately detect hydroxyl radicals and superoxide radicals.

3. Results and Discussion

The XRD pattern of the CdS nanoparticles used in this study is shown in Figure 1. According to the standard diffraction card PDF# 41-1049 of wurtzite CdS, all significant peaks can be indexed as those of wurtzite CdS, as shown in the figure. However, there is a very weak peak at 30.6°, which could be from an impurity phase.

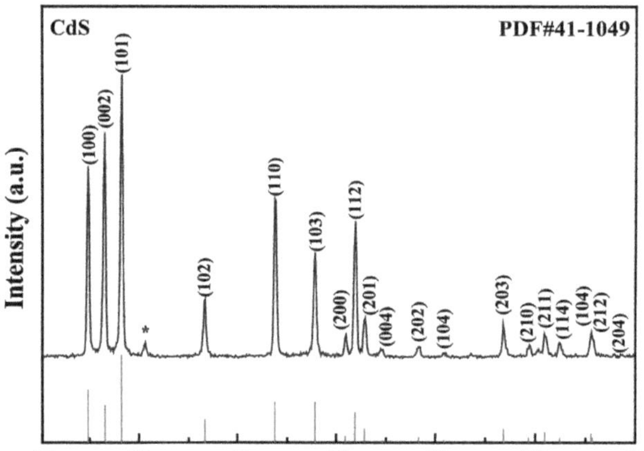

Figure 1. X-ray diffraction pattern of CdS powder used in this study.

Figure 2 presents the XP spectrum of the CdS nanoparticles. Besides the lines for Cd and S [37], only two lines for C 1s and O 1s, separately, can be observed. C impurity is well known for XPS analyses, while O 1s at 532.5 eV is most probably representative of chemisorbed oxygen [38]. Generally speaking, the CdS powder used in this study was of a high purity.

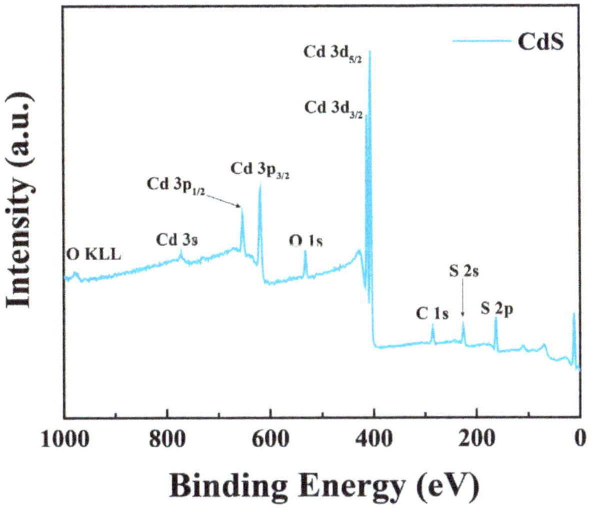

Figure 2. XP spectrum of CdS powder used in this study.

Figure 3 shows two representative SEM images obtained for the CdS powder in this study. From Figure 3a, it can be seen that the CdS particles are quite non-uniform in size, with large ones around 300 nm and small ones smaller than 50 nm. In addition, the CdS particles are of quite irregular shapes with clear edges, as shown in Figure 3b, which indicates a high degree of crystallinity [30].

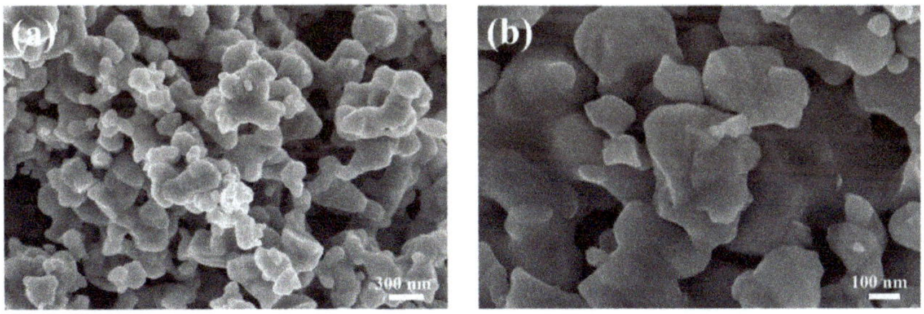

Figure 3. SEM images of CdS powder: (**a**) at a relatively small magnification; (**b**) at a larger magnification.

For the 50 mg/L RhB solution suspended with CdS in a glass-bottomed beaker, the absorption peak at 554 nm in the UV–VIS absorption spectrum decreased steadily with increasing stirring time, as shown in Figure 4a. The degradation efficiency of an organic dye is usually quantified using the formula $D = 1 - A/A_0$, where A_0 and A represent the initial and sustained intensity of the dye's characteristic absorption peak. After 8 h of magnetic stirring, the degradation efficiency was 78.9%. Due to its rather high initial concentration, the solution still exhibited a bright color, though its absorption peak had substantially decreased after 8 h of magnetic stirring, as shown in the inset of Figure 4a. In contrast, for the 50 mg/L RhB solution suspended with CdS in an Al_2O_3-bottomed beaker, after 8 h of magnetic stirring, the solution became colorless and its absorption peak at 554 nm in the UV–VIS absorption spectrum almost disappeared, as shown in Figure 4b. Obviously, the Al_2O_3 coating had imposed a remarkable enhancement on the tribocatalytic degradation of RhB by the CdS nanoparticles. Figure 4c compares the degradation efficiency of the RhB solution by CdS nanoparticles over time between the two kinds of beakers. On this basis, the pseudo-primary kinetics fit to the degradation efficiency of the RhB solution was further carried out, as shown in Figure 4d. The degradation rate constants of the CdS nanoparticles rubbed on glass and Al_2O_3 bottoms were 0.176 h^{-1} and 0.773 h^{-1}, respectively. The Al_2O_3 coating had increased the degradation rate constant of CdS nanoparticles by 4.38 times. It is worthy to mention that in a previous related investigation [36], 97.0% and 98.0% of 5 mg/L RhB were degraded by CdS nanowires after 14 h of magnetic stirring in a glass container and after 7 h of magnetic stirring in a polypropylene (PP) container, respectively. It is clear that PP also showed an enhancement on the tribocatalytic degradation of organic dyes by CdS, while its enhancement was smaller than that of Al_2O_3.

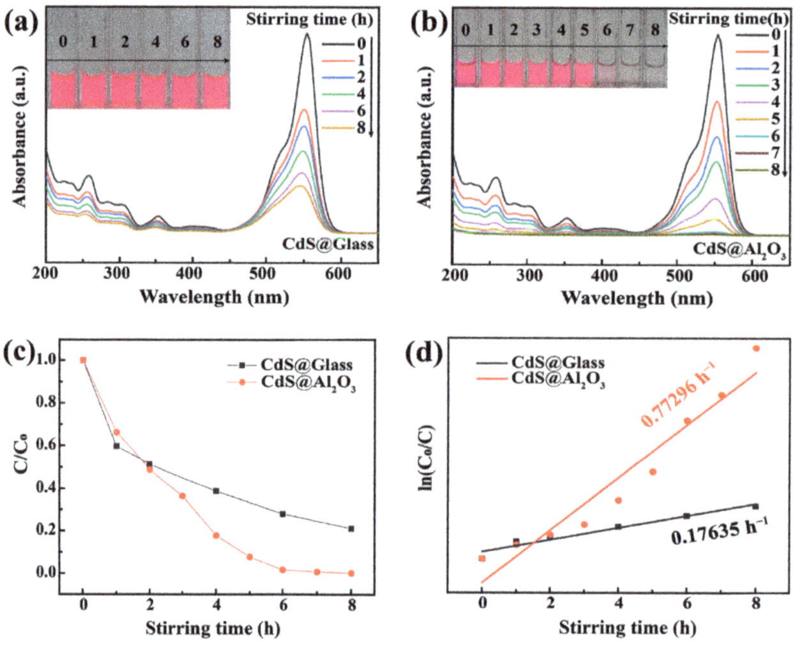

Figure 4. UV–VIS absorption spectra of RhB (50 mg/L) solutions mediated by CdS nanoparticles rubbing on different materials (Inset: color change in solutions): (**a**) glass; (**b**) Al_2O_3; (**c**) C/C_0 vs. stirring time under magnetic stirring with glass and Al_2O_3 bottoms; (**d**) kinetic curves.

For the 20 mg/L MO solution suspended with CdS in a glass-bottomed beaker, the absorption peak at 464 nm in the UV–VIS absorption spectrum decreased rather slowly with increasing stirring time, as shown in Figure 5a. After 24 h of magnetic stirring, the degradation efficiency was only 69.8% and the solution was still yellowish, as shown in the inset of Figure 5a. Moreover, as the absorption peak at 464 nm decreased, a new absorption peak at 250 nm appeared in the ultraviolet region. The same result had been observed in previous studies [39,40], which actually indicates that MO molecules were only broken into smaller organic molecules like benzoic acid, succinic acid, and p-phenol, according to mass spectrometry tests [28]. As a matter of fact, due to the presence of high-energy bonds (C=N and N=N) in its molecules, MO is relatively rather difficult to degrade among various common organic dyes, and this relatively slow degradation of 20 mg/L MO by CdS is rather usual for MO. To our great surprise, a different result was observed for the 20 mg/L MO solution suspended with CdS in the Al_2O_3-bottomed beaker. As shown in Figure 5b, after 12 h of magnetic stirring, the solution became colorless and its absorption peak at 464 nm in the UV–VIS absorption spectrum almost disappeared, and no new absorption peak was generated at 250 nm. This suggests that for the tribocatalytic degradation of MO by CdS nanoparticles, the Al_2O_3 coating not only significantly accelerated the speed, but also upgraded the degradation mode from a partial one to a complete one. Figure 5c compares the degradation efficiency of the MO solution by CdS nanoparticles over time between the two kinds of beakers, and Figure 5d presents the pseudo-primary kinetic fit to the degradation efficiency of the MO solution. The degradation rate constants of the CdS nanoparticles rubbed on the glass and Al_2O_3 coatings were 0.046 h^{-1} and 0.272 h^{-1}, respectively. The Al_2O_3 coating increased the degradation rate constant of CdS nanoparticles by 5.87 times.

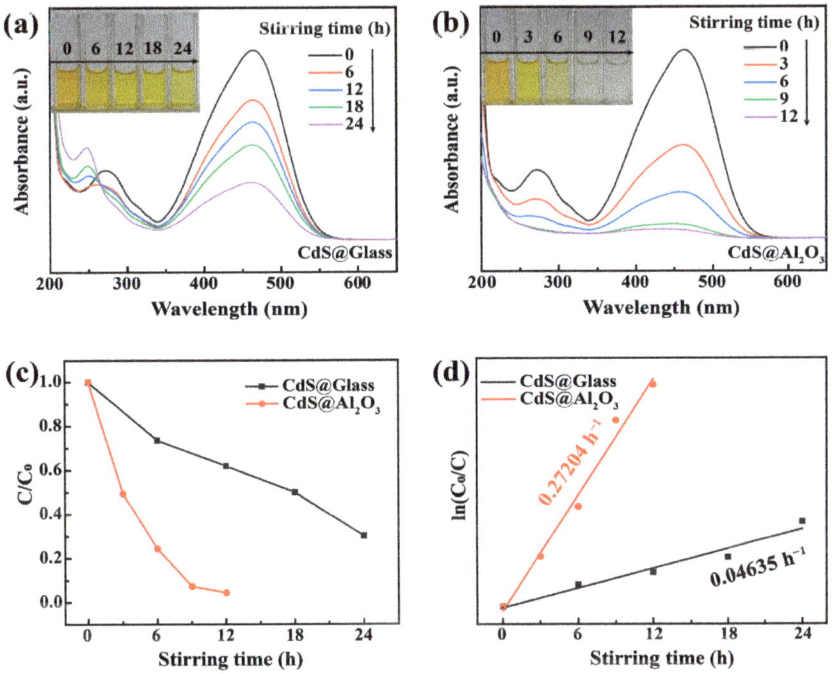

Figure 5. UV–VIS absorption spectra of MO (20 mg/L) solutions mediated by CdS nanoparticles rubbing on different materials (Inset: color change in solutions): (**a**) glass; (**b**) Al_2O_3; (**c**) C/C_0 vs. stirring time under magnetic stirring with glass and Al_2O_3 bottoms; (**d**) kinetic curves.

In order to examine the cycle stability of CdS, a cycle experiment of CdS nanoparticles was carried out, in which a 50 mg/L RhB solution suspended with CdS nanoparticles was degraded repeatedly via magnetic stirring. After 8 h of tribocatalysis, a small amount of a highly concentrated RhB dye solution was added to the degraded solution so that the solution concentration in the beaker was restored to 50 mg/L for a subsequent cycle. The degradation rate of the 50 mg/L RhB solution after 8 h of magnetic stirring was measured for every cycle, which was 99.9%, 100.0%, 99.9%, and 99.9% for the 1st, 2nd, 3rd, and 4th cycles, respectively. Obviously, CdS nanoparticles have an excellent cyclic stability and a great potential to degrade organic pollutants.

For both the tribocatalytic degradation of organic pollutants and the tribocatalytic conversion of H_2O and CO_2, coating materials on the bottoms of beakers/reactors has been found to be a convenient and effective method to enhance or regulate catalytic reactions [23,41–43]. The effects of the coatings are generally believed to result from the dynamic frictions between the coatings and the catalysts. As for the enhanced degradation of organic dyes observed for CdS nanoparticles in Al_2O_3-bottomed beakers in this study, the dynamic friction between Al_2O_3 and CdS nanoparticles must have played a vital role, as shown in a schematic drawing for the tribocatalytic process in Figure 6 [44].

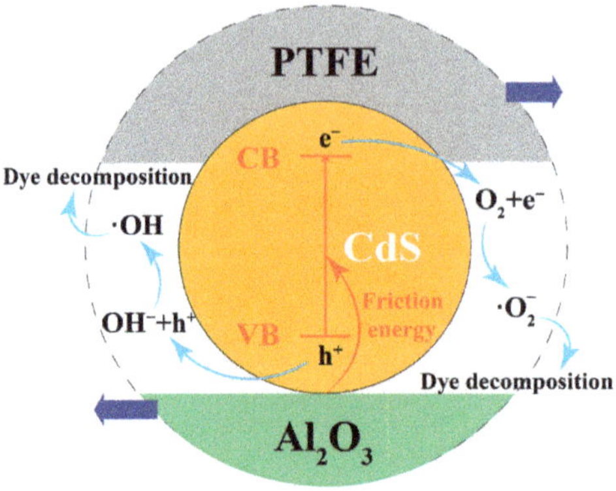

Figure 6. A schematic drawing for the enhanced tribocatalytic degradation of organic dyes by CdS nanoparticles in Al_2O_3-bottomed beakers.

According to the electronic transition mechanism for tribocatalysis [23], electron-hole pairs are excited in CdS by the mechanical energy absorbed through friction, which can be expressed as:

$$\text{CdS} \xrightarrow{\textit{Friction energy}} \text{CdS} + e^- + h^+ \tag{1}$$

Electrons further react with oxygen to form superoxide radicals, while holes react with hydroxide to form hydroxyl radicals, as follows:

$$OH^- + h^+ \rightarrow \cdot OH \tag{2}$$

$$O_2 + e^- \rightarrow \cdot O_2^- \tag{3}$$

Hydroxyl radicals and superoxide radicals further react with organic dyestuffs, in which organic dyestuffs are degraded and become pollution-free small molecules:

$$OH(or\ \cdot O_2^-) + Dye \rightarrow Decomposition \tag{4}$$

Hydroxyl radicals and superoxide radicals generated by CdS nanoparticles under magnetic stirring were detected through EPR, and the results are shown in Figure 7. Obviously, for CdS nanoparticles in beakers with both glass and Al_2O_3 bottoms, four distinct characteristic peaks of hydroxyl radicals with a ratio of (1:2:2:1) [45] were observed after the nanoparticles were stimulated via magnetic stirring for 15 min in deionized water (Figure 7a), and four characteristic peaks of superoxide radicals with a ratio of (1:1:1:1) [46] were detected after the nanoparticles were stimulated in methanol (Figure 7b). In addition, it can be seen that the peaks of both hydroxyl and superoxide radicals are stronger for the Al_2O_3-bottomed than for the glass-bottomed beaker, which would suggest that more radicals are generated for the Al_2O_3 ceramic bottom.

For catalysts in photocatalysis, various techniques have been adopted to modify them to improve their catalytic performances. For example, for titanium dioxide–bronze nanosheets, surface-exposed defect sites were formed through light illumination, which greatly enhanced their photocatalytic H_2 production rate [47]. For catalysts in tribocatalysis, similar techniques can also be adopted to reform them. For tribocatalysis, besides the modification of catalysts, coating materials on the bottoms of beakers is another convenient and effective method to enhance the catalytic effect. It is well known that Al_2O_3 ceramics are cheap, robust, and of extremely high chemical stability. Up to now, they have been employed as coatings in several tribocatalytic investigations and some highly surprising effects, including the ones observed in this study, have been revealed [24,43,48]. Presently, these effects cannot be satisfactorily understood. Further studies are highly desirable, which are important for the practical development and for achieving a better understanding of tribocatalysis.

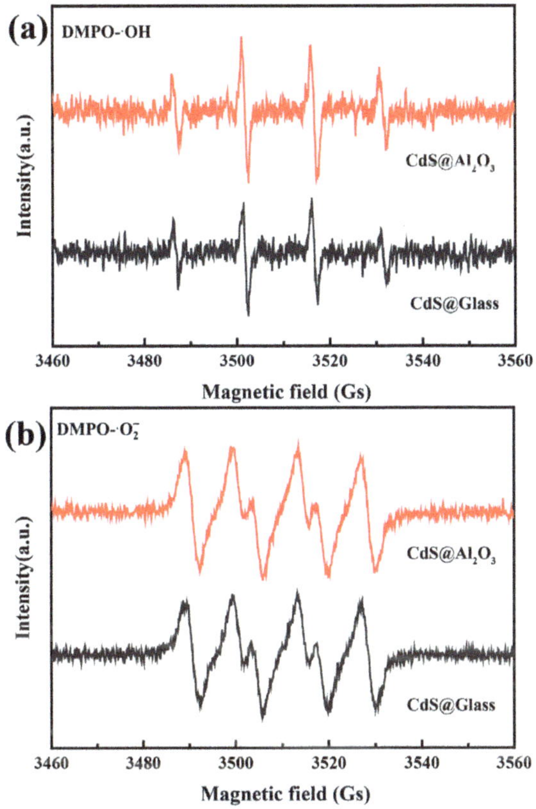

Figure 7. EPR spectra for CdS nanoparticles magnetically stirred in glass- and Al_2O_3-bottomed beakers containing: (**a**) deionized water with DMPO as the spin trapping agent; (**b**) methanol with DMPO as the spin trapping agent.

4. Conclusions

CdS nanoparticles have been employed to degrade 50 mg/L RhB and 20 mg/L MO solutions through magnetic stirring. With a Teflon magnetic rotary disk used to magnetically stir CdS nanoparticles in a glass beaker, 78.9% of the RhB and 69.8% of the MO solutions were degraded after 8 h and 24 h of magnetic stirring, respectively. With other conditions unchanged, the degradations were surprisingly enhanced through placing an Al_2O_3 disk on the beaker bottom—as much as 99.8% of the RhB and 95.6% of the MO solutions were degraded after 8 h and 12 h of magnetic stirring, respectively. In addition, for the degradation of MO by CdS nanoparticles in the glass beaker, a new peak appeared in the UV–visible absorption spectrum at 250 nm, indicating that the MO molecules are only decomposed into smaller organic molecules, such as benzoic acid, succinate, and p-phenol. However, with the Al_2O_3 coating, such a peak did not appear, indicating that the MO molecules are degraded mostly into H_2O and CO_2. These findings indicate that CdS has a great potential in the field of tribocatalysis and demonstrates that coating materials on vessel bottoms is a convenient and effective method for tribocatalysis enhancement.

Author Contributions: Conceptualization, S.K. and W.C.; methodology, S.K., C.M., R.L., and Z.Z.; validation, C.M., W.Z., and W.C.; formal analysis, S.K., C.M., R.L., and W.C.; investigation, S.K., C.M., R.L., and Z.Z.; data curation, C.M.; writing—original draft preparation, S.K.; writing—review and editing, C.M. and W.C.; visualization, C.M.; supervision, W.C.; project administration, Y.H., W.Z., and W.C.; funding acquisition, Y.H. and W.C. All authors have read and agreed to the published version of the manuscript.

Funding: This work was partially supported by the National Natural Science Foundation of China under Grant No. U21A20500, and the National Key R&D Program of China under Grant No. 2020YFB2008800.

Institutional Review Board Statement: Not applicable.

Informed Consent Statement: Not applicable.

Data Availability Statement: The data underlying this article will be shared on reasonable request to the corresponding author.

Conflicts of Interest: The authors declare no conflicts of interest.

References

1. Werber, J.R.; Osuji, C.O.; Elimelech, M. Materials for next-generation desalination and water purification membranes. *Nat. Rev. Mater.* **2016**, *1*, 16018. [CrossRef]
2. MacLeod, M.; Arp, H.P.H.; Tekman, M.B.; Jahnke, A. The global threat from plastic pollution. *Science* **2021**, *373*, 61–65. [CrossRef] [PubMed]
3. Cheng, L.; Xiang, Q.J.; Liao, Y.L.; Zhang, H.W. CdS-Based photocatalysts. *Energy Environ. Sci.* **2018**, *11*, 1362–1391. [CrossRef]
4. Lebreton, L.C.M.; van der Zwet, J.; Damsteeg, J.-W.; Slat, B.; Andrady, A.; Reisser, J. River plastic emissions to the world's oceans. *Nat. Commun.* **2017**, *8*, 15611. [CrossRef] [PubMed]
5. Rodriguez-Mozaz, S.; Chamorro, S.; Marti, E.; Huerta, B.; Gros, M.; Sànchez-Melsió, A.; Borrego, C.M.; Barceló, D.; Balcázar, J.L. Occurrence of antibiotics and antibiotic resistance genes in hospital and urban wastewaters and their impact on the receiving river. *Water Res.* **2015**, *69*, 234–242. [CrossRef] [PubMed]
6. Brillas, E.; Martínez-Huitle, C.A. Decontamination of wastewaters containing synthetic organic dyes by electrochemical methods. An updated review. *Appl. Catal. B Environ.* **2015**, *166–167*, 603–643. [CrossRef]
7. Cui, X.; Li, P.; Lei, H.; Tu, C.; Wang, D.; Wang, Z.; Chen, W. Greatly enhanced tribocatalytic degradation of organic pollutants by TiO_2 nanoparticles through efficiently harvesting mechanical energy. *Sep. Purif. Technol.* **2022**, *289*, 120814. [CrossRef]
8. Li, M.; Virguez, E.; Shan, R.; Tian, J.; Gao, S.; Patiño-Echeverri, D. High-resolution data shows China's wind and solar energy resources are enough to support a 2050 decarbonized electricity system. *Appl. Energy* **2022**, *306*, 117996. [CrossRef]
9. Daaboul, J.; Moriarty, P.; Honnery, D. Net green energy potential of solar photovoltaic and wind energy generation systems. *J. Clean. Prod.* **2023**, *415*, 137806. [CrossRef]
10. Aljundi, K.; Figueiredo, A.; Vieira, A.; Lapa, J.; Cardoso, R. Geothermal energy system application: From basic standard performance to sustainability reflection. *Renew. Energy* **2024**, *220*, 119612. [CrossRef]
11. Zhou, S.; Cao, S. Co-ordinations of ocean energy supported energy sharing between zero-emission cross-harbour buildings in the Greater Bay Area. *Appl. Energy* **2024**, *359*, 122718. [CrossRef]
12. Lei, H.; Wu, M.; Mo, F.; Ji, S.; Dong, X.; Wu, Z.; Gao, J.; Yang, Y.; Jia, Y. Tribo-catalytic degradation of organic pollutants through bismuth oxyiodate triboelectrically harvesting mechanical energy. *Nano Energy* **2020**, *78*, 105290. [CrossRef]
13. Wang, X.; Zhang, P.; Li, Z.; Zhang, S.; Ren, P.; Xu, J.; Hui, Y.; Dai, Z. Contrasting piezocatalytic and tribocatalytic behavior of $BaTiO_3$. *Mater. Sci. Semicond. Process.* **2024**, *172*, 108080. [CrossRef]
14. Ma, J.; Ren, J.; Jia, Y.; Wu, Z.; Chen, L.; Haugen, N.O.; Huang, H.; Liu, Y. High efficiency bi-harvesting light/vibration energy using piezoelectric zinc oxide nanorods for dye decomposition. *Nano Energy* **2019**, *62*, 376–383. [CrossRef]
15. Feng, R.; Xie, S.; Guan, W.; Zhong, Q. Tribocatalysis: Challenges and perspectives. *Sci. China Chem.* **2021**, *64*, 1609–1613.
16. Bie, C.; Wang, L.; Yu, J. Challenges for photocatalytic overall water splitting. *Chem* **2022**, *8*, 1567–1574. [CrossRef]
17. Li, Q.; Li, X.; Wageh, S.; Al-Ghamdi, A.A.; Yu, J.G. CdS/Graphene Nanocomposite Photocatalysts. *Adv. Energy Mater.* **2015**, *5*, 28. [CrossRef]
18. Xu, X.; Xiao, L.; Wu, Z.; Jia, Y.; Ye, X.; Wang, F.; Yuan, B.; Yu, Y.; Huang, H.; Zou, G. Harvesting vibration energy to piezo-catalytically generate hydrogen through Bi_2WO_6 layered-perovskite. *Nano Energy* **2020**, *78*, 105351. [CrossRef]

19. Kajdas, C.; Hiratsuka, K. Tribochemistry, tribocatalysis, and the negative-ion-radical action mechanism. *J. Eng. Tribol.* **2009**, *223*, 827–848. [CrossRef]
20. Song, W.; Li, J.F.; Zeng, C.Y.; Ouyang, C.K.; Sun, S.Y.; Wang, K.Q.; Li, J.J.; Luo, J.B. Tribo-catalysis triggered the in-situ formation of amphiphilic molecules to reduce friction and wear. *Tribol. Int.* **2023**, *185*, 10. [CrossRef]
21. Gao, K.; Bin, W.; Berman, D.; Ren, Y.; Luo, J.; Xie, G. Self-Adaptive Macroscale Superlubricity Based on the Tribocatalytic Properties of Partially Oxidized Black Phosphorus. *Nano Lett.* **2023**, *23*, 6823–6830. [CrossRef]
22. Li, P.C.; Wu, J.; Wu, Z.; Jia, Y.M.; Ma, J.P.; Chen, W.P.; Zhang, L.H.; Yang, J.; Liu, Y.S. Strong tribocatalytic dye decomposition through utilizing triboelectric energy of barium strontium titanate nanoparticles. *Nano Energy* **2019**, *63*, 103832. [CrossRef]
23. Lei, H.; Jia, X.; Wang, H.; Cui, X.; Jia, Y.; Fei, L.; Chen, W. Tribo-Catalytic Conversions of H_2O and CO_2 by NiO Particles in Reactors with Plastic and Metallic Coatings. *Coatings* **2023**, *13*, 396. [CrossRef]
24. Cui, X.D.; Wang, H.B.; Lei, H.; Jia, X.C.; Jiang, Y.; Fei, L.F.; Jia, Y.M.; Chen, W.P. Surprising Tribo-catalytic Conversion of H_2O and CO_2 into Flammable Gases utilizing Frictions of Copper in Water. *ChemistrySelect* **2023**, *8*, e202204146. [CrossRef]
25. Gaur, A.; Kumar Moharana, A.; Porwal, C.; Singh Chauhan, V.; Vaish, R. Degradation of organic dyes by utilizing $CaCu_3Ti_4O_{12}$ (CCTO) nanoparticles via tribocatalysis process. *J. Ind. Eng. Chem.* **2024**, *129*, 341–351. [CrossRef]
26. Gaur, A.; Porwal, C.; Chauhan, V.S.; Vaish, R. Tribocatalytic investigation of $BaTiO_3$ for dye removal from water. *J. Mater. Electron.* **2023**, *34*, 14. [CrossRef]
27. Mao, C.Y.; Lei, H.; Guo, Z.Y.; Jia, X.C.; Cui, X.D., Huang, J.W.; Fei, L.F.; Jia, Y.M.; Chen, W.P. Exceptional tribo-catalytic degradation of concentrated methyl orange and methylene blue solutions by DXN-RT30 TiO_2 nanoparticles. *Ceram. Int.* **2024**, *50*, 4737–4745. [CrossRef]
28. Cui, X.; Guo, Z.; Lei, H.; Jia, X.; Mao, C.; Ruan, L.; Zhou, X.; Wang, Z.; Chen, F.; Chen, W. Tribo-Catalytic Degradation of Methyl Orange Solutions Enhanced by Silicon Single Crystals. *Coatings* **2023**, *13*, 1804. [CrossRef]
29. Wu, M.; Zhang, Y.; Yi, Y.; Zhou, B.; Sun, P.; Dong, X. Regulation of friction pair to promote conversion of mechanical energy to chemical energy on Bi_2WO_6 and realization of enhanced tribocatalytic activity to degrade different pollutants. *J. Hazard. Mater.* **2023**, *459*, 132147. [CrossRef] [PubMed]
30. Lei, H.; Cui, X.; Jia, X.; Qi, J.; Wang, Z.; Chen, W. Enhanced Tribocatalytic Degradation of Organic Pollutants by ZnO Nanoparticles of High Crystallinity. *Nanomaterials* **2023**, *13*, 46. [CrossRef]
31. Cao, J.; Jia, Y.; Wan, X.; Li, B.; Zhang, Y.; Huang, S.; Yang, H.; Yuan, G.; Li, G.; Cui, X.; et al. Strong tribocatalysis of strontium titanate nanofibers through harvesting friction energy for dye decomposition. *Ceram. Int.* **2022**, *48*, 9651–9657. [CrossRef]
32. Zhang, Q.; Jia, Y.; Wang, X.; Zhang, L.; Yuan, G.; Wu, Z. Efficient tribocatalysis of magnetically recyclable cobalt ferrite nanoparticles through harvesting friction energy. *Sep. Purif. Technol.* **2023**, *307*, 122846. [CrossRef]
33. Jie, L.F.; Gao, X.; Cao, X.Q.; Wu, S.; Long, X.X.; Ma, Q.Y.; Su, J.X. A review of CdS photocatalytic nanomaterials: Morphology, synthesis methods, and applications. *Mater. Sci. Semicond. Process.* **2024**, *176*, 18. [CrossRef]
34. Li, X.; Zhu, J.; Li, H. Comparative study on the mechanism in photocatalytic degradation of different-type organic dyes on SnS_2 and CdS. *Appl. Catal. B Environ.* **2012**, *123–124*, 174–181. [CrossRef]
35. Fard, N.E.; Fazaeli, R.; Ghiasi, R. Band Gap Energies and Photocatalytic Properties of CdS and Ag/CdS Nanoparticles for Azo Dye Degradation. *Chem. Eng. Technol.* **2016**, *39*, 149–157. [CrossRef]
36. Yang, B.A.; Chen, H.B.; Guo, X.D.; Wang, L.; Xu, T.; Bian, J.H.; Yang, Y.D.; Liu, Q.D.; Du, Y.P.; Lou, X.J. Enhanced tribocatalytic degradation using piezoelectric CdS nanowires for efficient water remediation. *J. Mater. Chem. C* **2020**, *8*, 14845–14854. [CrossRef]
37. Stoev, M.; Katerski, A. XPS and XRD study of photoconductive CdS films obtained by a chemical bath deposition process. *J. Mater. Chem.* **1996**, *6*, 377–380. [CrossRef]
38. Song, J.; Zhao, J.; Liu, Y.; Hu, Y.; Chen, W. Room-temperature hydrogen sensitive Pt—SnO_2 composite nanoceramics: Dormancy and a practicable regeneration method. *Ceram. Int.* **2024**, *50*, 31357–31363. [CrossRef]
39. Wang, M.; Li, M.; Xu, L.; Wang, L.; Ju, Z.; Li, G.; Qian, Y. High yield synthesis of novel boron nitride submicro-boxes and their photocatalytic application under visible light irradiation. *Catal. Sci. Technol.* **2011**, *1*, 1159–1165. [CrossRef]
40. Filice, S.; D'Angelo, D.; Libertino, S.; Nicotera, I.; Kosma, V.; Privitera, V.; Scalese, S. Graphene oxide and titania hybrid Nafion membranes for efficient removal of methyl orange dye from water. *Carbon* **2015**, *82*, 489–499. [CrossRef]
41. Jia, X.C.; Wang, H.B.; Lei, H.; Mao, C.Y.; Cui, X.D.; Liu, Y.; Jia, Y.M.; Yao, W.Q.; Chen, W.P. Boosting tribo-catalytic conversion of H_2O and CO_2 by Co_3O_4 nanoparticles through metallic coatings in reactors. *J. Adv. Ceram.* **2023**, *12*, 1833–1843. [CrossRef]
42. Mao, C.Y.; Zhang, Y.C.; Lei, H.; Jia, X.C.; Chen, F.; Yao, W.Q.; Liu, P.T.; Chen, W.P. Boosting tribo-catalytic degradation of organic pollutants by $BaTiO_3$ nanoparticles through metallic coatings. *Appl. Surf. Sci.* **2024**, *663*, 160172. [CrossRef]
43. Xu, X.; Mao, C.; Song, J.; Ke, S.; Hu, Y.; Chen, W.; Pan, C. Surprising Effects of Ti and Al_2O_3 Coatings on Tribocatalytic Degradation of Organic Dyes by GaN Nanoparticles. *Materials* **2024**, *17*, 3487. [CrossRef] [PubMed]
44. Wang, Z.M.; Berbille, A.; Feng, Y.W.; Li, S.; Zhu, L.P.; Tang, W.; Wang, Z.L. Contact-electro-catalysis for the degradation of organic pollutants using pristine dielectric powders. *Nat. Commun.* **2022**, *13*, 9. [CrossRef] [PubMed]
45. Lin, F.; Zhang, Y.; Wang, L.; Zhang, Y.; Wang, D.; Yang, M.; Yang, J.; Zhang, B.; Jiang, Z.; Li, C. Highly efficient photocatalytic oxidation of sulfur-containing organic compounds and dyes on TiO_2 with dual cocatalysts Pt and RuO_2. *Appl. Catal. B Environ.* **2012**, *127*, 363–370. [CrossRef]
46. Duan, Y.; Luo, J.; Zhou, S.; Mao, X.; Shah, M.W.; Wang, F.; Chen, Z.; Wang, C. TiO_2-supported Ag nanoclusters with enhanced visible light activity for the photocatalytic removal of NO. *Appl. Catal. B Environ.* **2018**, *234*, 206–212. [CrossRef]

47. Rej, S.; Hejazi, S.M.H.; Badura, Z.; Zoppellaro, G.; Kalytchuk, S.; Kment, Š.; Fornasiero, P.; Naldoni, A. Light-Induced Defect Formation and Pt Single Atoms Synergistically Boost Photocatalytic H_2 Production in 2D TiO_2-Bronze Nanosheets. *ACS Sustain. Chem. Eng.* **2022**, *10*, 17286–17296. [CrossRef]
48. Lei, H.; Wu, Z.; Wang, H.; Mao, C.; Guo, Z.; Fei, L.; Chen, W. Converting H_2O and CO_2 into chemical fuels by nickel via friction. *Surf. Interfaces* **2024**, *46*, 104203. [CrossRef]

coatings

MDPI

Article

Photocatalytic Degradation of Sulfamethoxazole by Cd/Er-Doped Bi_2MoO_6

Nengxun Yang *,†, Yixuan Niu †, Bohang Zhang and Fuchun Zhang

School of Physics and Electronic Information, Yan'an University, Yan'an 716000, China; 15091006586@163.com (Y.N.); zhangbohang9813@163.com (B.Z.); yadxzfc@yau.edu.cn (F.Z.)
* Correspondence: yadxynx@126.com
† These authors contributed equally to this work.

Abstract: Bi_2MoO_6 (BMO) is a typical bismuth-based semiconductor material, and its unique Aurivillius structure provides a broad space for electron delocalization. In this study, a new type of bismuth molybdate Cd/Er-BMO photocatalytic material was prepared by co-doping Er^{3+} and Cd^{2+}, and the performance of the photocatalytic degradation of sulfamethoxazole (SMZ) was systematically studied. The research results showed that the efficiency of SMZ degradation by Cd/Er-BMO was significantly improved after doping Er^{3+} and Cd^{2+} ions, reflecting the synergistic catalytic effect of Cd^{2+} and Er^{3+} co-doping. Cd/Er-BMO doped with 6% Cd had the highest degradation efficiency (93.89%) of SMZ under visible light irradiation. The material revealed excellent stability and reusability in repeated degradation experiments. In addition, 6% Cd/Er-BMO had a smaller particle size and a larger specific surface area, which is conducive to improving the generation efficiency of its photogenerated electron-hole pairs and reducing the recombination rate, significantly enhancing the photocatalysis of the material. This study not only provides an effective photocatalyst for degrading environmental pollutants such as SMZ, but also provides an important scientific basis and new ideas for the future development of efficient and stable photocatalytic materials.

Keywords: photocatalysis; Bi_2MoO_6; hydrothermal method; co-doping

1. Introduction

With the development of global industrialization, the human living environment has gradually deteriorated, and a large amount of waste has been discharged. Water pollution is particularly prominent among many pollution problems as it contains a large amount of organic pollutants, many of which are difficult to degrade, such as phenols, polychlorinated biphenyls, and polycyclic aromatic hydrocarbons. These organic pollutants possess a relatively high biological toxicity, which seriously threatens human health and life. According to the survey, a large number of additives that pollute the environment and are harmful to the human body are used in the process of textile printing and dyeing. Most of these additives are discharged into the water environment, resulting in water pollution. In addition, upon the increase in the use of antibiotics in human society, antibiotics inevitably enter the water environment. Among them, sulfamethoxazole (SMZ) is an organic compound that is mainly used to treat diseases such as urinary tract infections and respiratory system infections that are caused by sensitive bacteria. It has shown strong antibacterial properties in clinical practise. However, studies have shown that 45%–90% will be directly excreted in the form of metabolites through urine and feces and enter the environment [1]. Conventional methods, such as physical adsorption and Fenton methods, are difficult to effectively degrade these compounds, leading to the long-term deterioration of water quality, which, in turn, affects the virtuous cycle of the ecosystem. Moreover, sulfonamides are degraded slowly after entering the environment and are likely to remain for a long time, which will eventually result in an adverse impact on human health. As such, the degradation treatment of this type of wastewater is of great importance [2].

In order to remove harmful pollutants from the water environment, photocatalytic degradation was adopted in this work. Photocatalysis is a pollutant degradation technology that makes use of radiation, usually in the UV and visible range, in the presence of an adequate catalyst to produce highly reactive oxidizing species. It is a promising environmental technology due to its high efficiency, environmental friendliness, and energy saving characteristics. It can remove reluctant-to-degrade organic pollutants in wastewater at a low cost [3]. This method can decompose organic substances that are harmful to the human body and the environment without causing a waste of resources or producing secondary pollutants. A large number of studies have shown that organic pollutants can be effectively photocatalytically degraded, decolorized, and mineralized into small inorganic molecules, thereby eliminating pollution to the environment. Photocatalysts have also attracted much attention given their strong redox ability, good stability, low pollution, and easy recycling properties. UV-excited TiO_2-based photocatalysts have been widely studied [4]. The principle of this technology is to use TiO_2 to generate active free radicals, thereby achieving the photocatalytic degradation of organic pollutants and generating CO_2, H_2O, and simple inorganic substances. However, as a result of the wide band gap of TiO_2 (3.2 eV), it can only absorb ultraviolet light and does not have good catalytic activity in the visible light range, which greatly limits the practical application of such catalysts. Studies have shown that the properties of titanium dioxide can be changed by making composite materials [5–7]. In addition, in-depth research on semiconductor materials such as CdS [8,9], ZrO_2 [10,11], and SnO_2 [12,13] has been widely conducted, and has achieved remarkable results. However, the photogenerated electrons and holes in CdS are easy to recombine, which reduces its photocatalytic efficiency. The wide band gap energy of ZrO_2 results in less light absorption and its photocatalytic activity is limited by the high charge carrier recombination efficiency. The inherent defects of SnO_2 in photocatalysis, such as structural design and morphology control, have been widely documented, which can affect its photocatalytic performance.

In view of the above shortcomings of photocatalytic materials, more scholars have begun to study new ones in recent years. Bismuth-based semiconductor materials have attracted extensive attention worldwide as a result of their easy regulatory morphology, good photochemical stability, and unique electronic band structure [14]. Bismuth molybdate (Bi_2MoO_6) is a compound with remarkable photocatalytic properties and is one of the most important members of the Aurivillius oxide materials family [15], with its structure providing sufficient space for electron delocalization, thereby facilitating catalytic reactions [16]. However, the practical application of the Bi_2MoO_6 monomer is severely limited given its poor photocarrier separation, high recombination rate, and narrow photoresponse range [17]. To overcome these limitations, a variety of modifications have been made in recent years. Current modification methods mainly focus on doping with non-metallic ions, transition metals, and rare earth metals [18]. Wang et al. [19] synthesized Cu-doped Bi_2MoO_6 using a simple solvothermal method. Cu doping reduces the work function and improves the charge separation efficiency, which is considered to be the main reason for the enhanced photoactivity. In addition, Cu doping has little effect on the morphology of Bi_2MoO_6, though it has a greater effect on the energy band structure, which makes the reducibility of Cu-doped Bi_2MoO_6 stronger. As such, Cu-Bi_2MoO_6 exhibits a higher photocatalytic efficiency than pure Bi_2MoO_6. Meng et al. [20] prepared an efficient Bi_2MoO_6 nitrogen-fixing photocatalyst using iron as the medium. The Fe-induced reduction in the surface work function facilitates charge transport to the catalyst surface. In addition, Fe doping can also improve charge collection through the Fe^{3+}/Fe^{2+} redox pathway and can become an active site for nitrogen reduction. Given the above advantages, Fe-mediated Bi_2MoO_6 significantly enhances the photocatalytic activity of nitrogen fixation driven by visible light. Li et al. [21] prepared Eu^{3+}-doped bismuth molybdate phosphor using the sol–gel method, studied the effect of different calcination temperatures on BMO: Eu^{3+}, and characterized its structure and properties. The crystal structure and related parameters were obtained using the Rietveld method. The relationship between crystal structure and luminescence properties was analyzed in detail, and BMO: Eu^{3+} phosphor was applied to the preparation of luminescent films and LED devices. As such, it is very important to select a suitable doping atom to improve the photocatalytic degradation efficiency.

Previous studies have shown that sulfamethoxazole and rhodamine B can be photocatalytically degraded under visible light by doping γ-Bi_2MoO_6 nanomaterials with heterovalent cadmium. Researchers have successfully synthesized a Cd-BMO photocatalyst with an excellent pollutant degradation efficiency through a simple hydrothermal method [18,22]. Compared with BMO, Cd-BMO has a smaller particle size, a larger specific surface area, a greater charge separation efficiency, and a greater electron excitation capability, thereby achieving a higher degradability of pollutants. The latest research indicates that the co-doping of multiple elements can lead to more effective photocatalysts through the synergistic interaction between multiple ions. For example, Li et al. [23] used a hydrothermal method to synthesize a new type of Gd/Er/Lu-doped Bi_2MoO_6 photocatalyst for the degradation of rhodamine B (RhB) and tetrachlorophenol, significantly enhancing the photocatalytic degradation performance. Specifically, after doping with Gd^{3+}, hydroxyl radicals were generated to improve the oxidation efficiency of Bi_2MoO_6. The introduction of Er^{3+} provided an energy upconversion centre, thereby improving light absorption. After doping with Lu^{3+} ions, abundant oxygen vacancies were generated in the Bi_2MoO_6 crystal, thereby promoting carrier separation. Wang et al. [24] used the citric acid complexation method to prepare Eu^{3+} and Fe^{3+} co-doped bismuth molybdate to improve its photocatalytic performance. The results indicated that Eu^{3+} and Fe^{3+} co-doped bismuth molybdate achieved the highest photocatalytic activity. The synergistic effect of the two ions resulted in a 94.1% degradation of the sample within 50 min, and the photocatalytic activity of Eu/Fe-BMO was not simply the sum of the photocatalytic properties of the Eu-BMO and Fe-BMO catalysts. Inspired by the synergistic effect of co-doping, this study used a hydrothermal method to prepare Er^{3+} and Cd^{2+} co-doped photocatalytic materials. Through material characterization and photocatalytic degradation experiments, the synergistic effect of the two doping ions was studied to further modify the intrinsic material and improve its photocatalytic performance.

2. Materials and Methods

2.1. The Preparation of the Catalyst

Er-BMO and Cd/Er-BMO materials were prepared using the hydrothermal method. The chemical reagents used in the preparation process were not purified. The detailed preparation process is as follows: $Bi(NO_3)_3{\cdot}5H_2O$ (4 mmol) was dissolved in 10 mL HNO_3 (2 $mol{\cdot}L^{-1}$) to obtain solution A, and $(NH_4)_6Mo_7O_{24}{\cdot}4H_2O$ (0.2857 mmol) was dissolved in 10 mL NaOH (1 $mol{\cdot}L^{-1}$) to obtain solution B. Solution B was slowly added to solution A and was stirred continuously for 20 min to obtain solution C, before a certain amount of $Er(NO_3)_3{\cdot}5H_2O$ (0.12 mmol) was added. Then, solid $Cd(NO_3)_2{\cdot}4H_2O$ (0.08, 0.16, 0.24, 0.32, and 0.4 mmol) was weighed and added while continuously stirring until uniform. The pH of the solution was adjusted to 8 with ammonia solution. Finally, the precursor solution was transferred to a 100 mL polytetrafluoroethylene liner, kept at 180 °C for 12 h, and was cooled naturally to room temperature. The samples were washed with deionized water and anhydrous ethanol, respectively, and were dried at 80 °C for 12 h. The samples were labelled as X Cd/Er-BMO (X = 2%, 4%, 6%, 8%, and 10%). In addition, the γ-Bi_2MoO_6 sample containing only Er^{3+} was prepared using the same method and was labelled as Er-BMO.

2.2. Characterization

The crystal structure of the samples was analyzed using X-ray diffraction (XRD), in the range of 5–90°, at a scanning rate of 2°/min and a step size of 0.02°. The samples were studied using scanning electron microscopy (SEM) at a working voltage of 15 kV to reveal the morphology of the samples. The microstructure of the samples was analyzed using transmission electron microscopy (TEM) at 250 kV. The specific surface area of the samples was determined using the Bruauer-Emmett-Teller (BET) method, and the average pore size of the samples was determined using nitrogen adsorption at 77 K. The samples were degassed under vacuum conditions at 150 °C for 6 h. The instrument used in the measurement process was Quanta 4000 (Quanta Computer, Taoyuan, China). The elemental valence state of the samples was analyzed using X-ray photoelectron spectroscopy (XPS) using Moser K-ALPHA (Thermo Fisher Scientific, Waltham, MA, USA). The prepared samples were measured using ultraviolet-visible diffuse reflectance spectroscopy (UV-Vis DRS) using a UV/VIS/NIR Lambda 1050 (PerkinElmer, Waltham, MA, USA). The optical response and electron transfer characteristics of the sample were analyzed by using the transient photocurrent response. Among them, the suspension prepared using Na_2SO_4 was used for photocurrent tests, and the samples were tested under a 300 W xenon light source and electrochemical workstation. Photoluminescence (PL) was used to analyze the recombination rate of photogenerated electron-hole pairs, and the time-resolved fluorescence data of the samples were measured using an Hitachi F-7100 (Hitachi High-Technologies, Tokyo, Japan). The photoresponse ability of the prepared samples was analyzed using the transient photocurrent response.

2.3. Photocatalytic Experiment

The prepared sample was placed under a long arc xenon lamp to study its photocatalytic degradation performance of the SMZ solution. A total of 50 mL of SMZ pollutant solution (5 mg·L^{-1}) and 50 mg of catalyst powder were placed in a quartz tube and were irradiated with a 400 W xenon lamp. The solution was stirred continuously for 30 min without visible light irradiation, and part of the suspension was taken after reaching the adsorption-desorption equilibrium. The xenon lamp was turned on and a certain amount of suspension (SMZ solution) was withdrawn every 45 min. It is important to use a cooling jacket to control the temperature of the photocatalytic reaction at 15 °C. Finally, the pollutant concentration was determined spectrophotometrically in the range of 190–338 nm using a UV1901PC instrument (Unico (Shanghai) Instruments Co., Ltd., Shanghai, China).

3. Results and Discussion

3.1. Characterization of Photocatalysts

X-ray diffraction (XRD) was used to analyze and confirm the crystal structure and phase composition of the existing catalyst (Figure 1a). As clearly shown in the figure, the XRD pattern of the Cd/Er-BMO sample is very similar to the reference pattern JCPDS No. 21-0102 of the original BMO; the diffraction peak is sharp, indicating that the crystallinity of the prepared sample is good [25]. In the Cd/Er-BMO sample, the 2θ diffraction angles are 11.0°, 27.2°, 28.3°, 32.5°, 33.2°, 36.1°, 46.7°, 47.1°, 55.6°, and 56.4°, respectively, corresponding to the PDF card of γ-Bi_2MoO_6 (JCPDS No. 21-0102) of (020), (140), (131), (200), (060), (151), (202), (260), (133), and (082). Among them, the most significant diffraction peak of the sample was observed at the (131) crystal plane. Upon the addition of different concentrations of Cd^{2+}, the peak at (131) continued to change but no other diffraction peaks were generated, further illustrating the successful doping of Cd^{2+} into Er-BMO without producing other substances [20]. Meanwhile, the resolution of the diffraction peak of the sample was high, indicating that the crystallinity was good. In particular, note that under different doping ratios, almost no additional diffraction peaks were revealed in the XRD images of Cd/Er-BMO, indicating that the crystal structure of Er-BMO was not changed after the introduction of Cd^{2+}. The results showed that compared with Er-BMO, the (131) diffraction peak of Cd/Er-BMO gradually increased with the increase in Cd^{2+} doping concentration, and the highest peak appeared in the Cd/Er-BMO sample when Cd was 6%. The peak intensity increased and the shape of the diffraction peak was clear, indicating that the formation of nanocrystals was good. However, the subsequent peak drop can be explained by the excessive Cd^{2+} doping concentration inhibiting the growth of crystal size, resulting in a smaller crystal size [26]. Figure 1b shows that the Cd/Er-BMO (131) diffraction peak shifts to larger diffraction angles with increasing Cd^{2+} doping concentration and reaches a maximum of 28.42° at 6% Cd/Er-BMO. The shift of the diffraction peak reflects the contraction of the crystal and the reduction in the crystal plane spacing [27]. This is because the small ionic radius element replaces the large ionic radius element, because the Bi^{3+} ionic radius (1.08 Å) is larger than the Cd^{2+} ionic radius (0.97 Å). The subsequent shift to the left is due to lattice distortion caused by the high Cd^{2+} doping concentration [18].

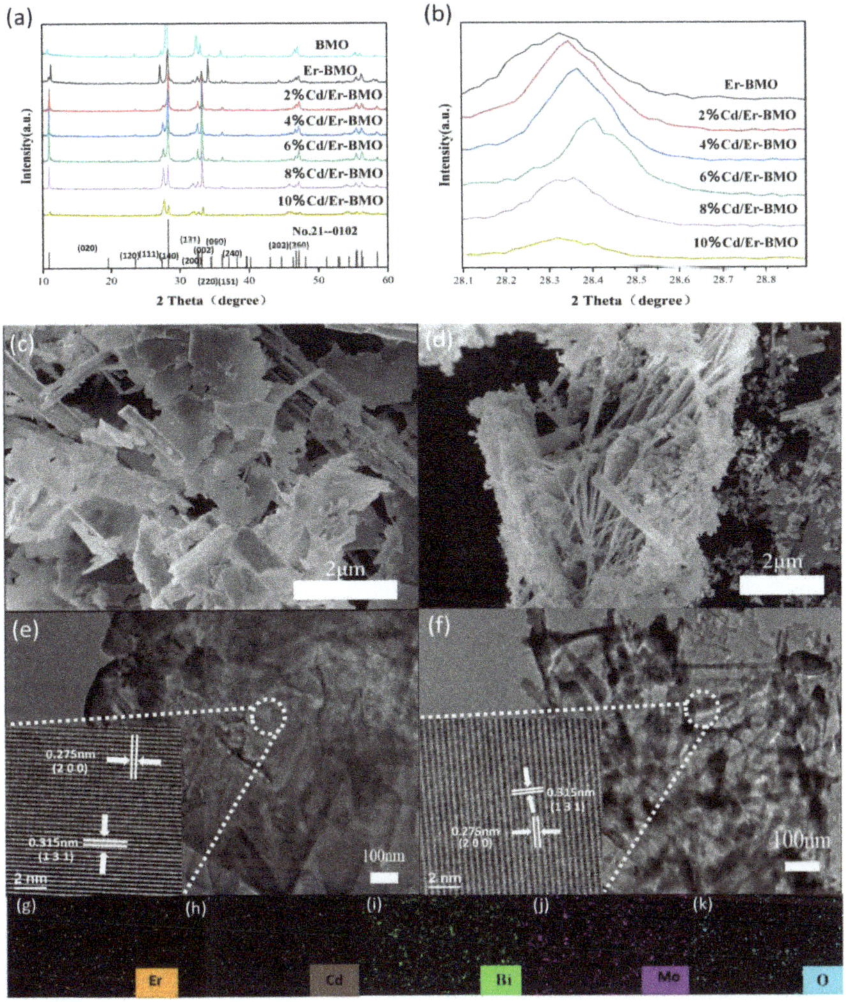

Figure 1. (**a**) XRD patterns of BMO, Er-BMO, and Cd/Er-BMO; (**b**) XRD patterns containing the (131) diffraction peak; (**c**) SEM image of Er-BMO; (**d**) SEM image of 6% Cd/Er-BMO; (**e**) TEM and HRTEM images of Er-BMO; (**f**) TEM and HRTEM images of 6% Cd/Er-BMO; (**g–k**) EDS elemental map of 6% Cd/Er-BMO.

The morphology and particle size of the synthesized samples were observed and studied using scanning electron microscopy (SEM). No particles, cracks, or surface roughness changes were observed in Er-BMO (Figure 1c) and 6% Cd/Er-BMO (Figure 1d). Compared with Er-BMO materials, the size of the 6% Cd/Er-BMO nanosheets was significantly reduced, changing from the original sheet shape to the needle shape. The emergence of the needle-like structure indicated that the addition of Cd^{2+} resulted in a significant increase in the specific surface area of the intrinsic material and an increase in the contact area for the reaction. The appearance of the needle-like structure may be attributable to the doping of Cd^{2+} ions, which affected the growth of Er-BMO [28].

The microstructure and lattice planes of Er-BMO and 6% Cd/Er-BMO were measured using transmission electron microscopy (TEM) and high-resolution transmission electron microscopy (HRTEM) (Figure 1e,f) to investigate the microstructure and properties of the materials. Both the Er-BMO and 6% Cd/Er-BMO samples have complete morphology, changing from an irregular flake structure to a needle-like structure. It is worth noting that the size of 6% Cd/Er-BMO was smaller than that of Er-BMO, which is consistent with the SEM analysis. HRTEM images showed the lattice features of Er-BMO (Figure 1e) and 6% Cd/Er-BMO (Figure 1f). The lattice fringes of Er-BMO and Cd/Er-BMO were clear, corresponding to the (131) and (200) planes of the orthorhombic crystal of γ-Bi_2MoO_6, respectively, indicating that there was no change in the orthorhombic phase before and after doping. The TEM and HRTEM images confirmed that the doping of Cd^{2+} had no significant effect on the crystal lattice and did not produce any other crystalline phase, which is consistent with the XRD results mentioned above [23].

The specific surface area is an important parameter in the photocatalytic degradation process that affects the performance and efficiency of the catalyst. A larger specific surface area can increase the active sites, improve the adsorption capacity, and increase light absorption. The specific surface area (BET) of Er-BMO and 6% Cd/Er-BMO was determined using N_2 adsorption/desorption isotherms. The pore size distribution was determined using the Barrett-Joyner-Halenda method (BJH). As shown in Figure 2a,b, in the relative pressure range of 0.8 to 1.0, the isotherms of the two samples showed typical type IV characteristics, namely, the adsorption amount was caused by capillary condensation [29] and formed a hysteresis loop in the medium and high relative pressure region. The figure shows that the 6% Cd/Er-BMO crystals had a stronger adsorption/desorption curve. Figure 2b reveals that the pore size distribution of the sample was between 2 and 50 nm. This pore size is between micropores and macropores, which belong to a mesoporous structure. Mesoporous materials usually have a high specific surface area and good fluid transport properties. The specific surface area and average pore size of the tested samples are shown in the insert table in Figure 2. When doping with Cd^{2+}, the specific surface area of Er-BMO increased from 8.414 to 12.638 $m^2 \cdot g^{-1}$, indicating that the doping of Cd^{2+} ions resulted in a larger specific surface area of Er-BMO, as well as more active sites. Meanwhile, the increase in active sites equates to more catalytic active centres, which can effectively improve the rate and efficiency of the catalytic reaction. This is consistent with the above SEM analysis results [30]. In addition, the study also showed that the average pore size of Cd/Er-BMO was better than that of Er-BMO, which resulted in a larger number of active surface sites [31].

An energy spectrum (EDS) element diagram of 6% Cd/Er-BMO (Figure 1g–k) shows that the Cd, Er, Bi, Mo, and O elements are uniformly distributed, which means that Cd^{2+} ions are successfully doped into Er-BMO.

Analyzing the chemical state of the doped sample benefits the study of the bonding of elements. As such, XPS analysis was used to monitor the valence state and elemental composition of Er-BMO and Cd/Er-BMO. Prior to the analysis of the data, C 1s was used for calibration at 284.8 eV (Figure 2c) given its stability, universality, clear peaks, and simple electronic structure. It can be observed that the characteristic peaks of the C, Cd, Er, Bi, Mo, and O atoms appeared in the XPS spectrum of 6% Cd/Er-BMO. Among them, the presence of the C element is attributed to the surface contamination of carbon materials (Figure 2d) [32], and the peak at 288.4 eV may be attributable to C-N or C-$(N)_3$ in the aromatic lattice [33]. As shown in Figure 2e, Cd 3d is located at 405.0 and 411.6 eV, corresponding to Cd $3d_{5/2}$ and Cd $3d_{3/2}$ of Cd^{2+}, respectively. Two strong peaks at 158.9 and 164.3 eV were observed in the Bi 4f XPS spectrum of 6% Cd/Er-BMO, which corresponds to Bi $4f_{7/2}$ and Bi $4f_{5/2}$, respectively (Figure 2f), indicating that the valence state of Bi in 6% Cd/Er-BMO is +3 [18]. Similarly, Mo 3d corresponded to the two peaks of Mo $3d_{5/2}$ and

Mo $3d_{3/2}$ that were located at 232.2 and 235.3 eV, respectively (Figure 2g), indicating that Mo exists only in the Mo^{6+} oxidation state in 6% Cd/Er-BMO [34]. Two strong peaks at 170.2 and 169.0 eV were shown in the spectrum of Er 4d, corresponding to Er $4d_{3/2}$ and Er $4d_{5/2}$, respectively (Figure 2h). The O 1s peaks of Er-BMO and 6% Cd/Er-BMO are shown in Figure 2i. The O 1s peak was fitted using three Gaussian superposition peaks, corresponding to 530.0, 531.2, and 532.7 eV, respectively. The peak at 530 corresponds to lattice oxide species, while the peak at 531.2 corresponds to surface hydroxyl groups and the peak at 532.6 corresponds to weakly adsorbed water [35]. The comparison result regarding the peaks of Er-BMO and Cd/Er-BMO indicated that doping Cd^{2+} has no effect on the valence state of Er-BMO. Cd/Er-BMO has similar peaks to Er-BMO, indicating that their crystal structures are similar. The reduction in electron density weakens the electron shielding effect, thereby affecting the binding energy and bonding environment. After doping, the Gaussian peak of the Er-BMO sample shifts toward the high bond energy direction, indicating that doping with Cd^{2+} will change the chemical environment [36], replacing Bi^{3+} or Er^{3+} to form ionic bonds with oxygen atoms, or Cd^{2+} is inserted into the Er-BMO lattice, resulting in a decrease in the electron density of Bi^{3+}, Mo^{6+}, and O^{2-} [37]. This also proves that Cd is successfully incorporated into the Er-BMO lattice [38].

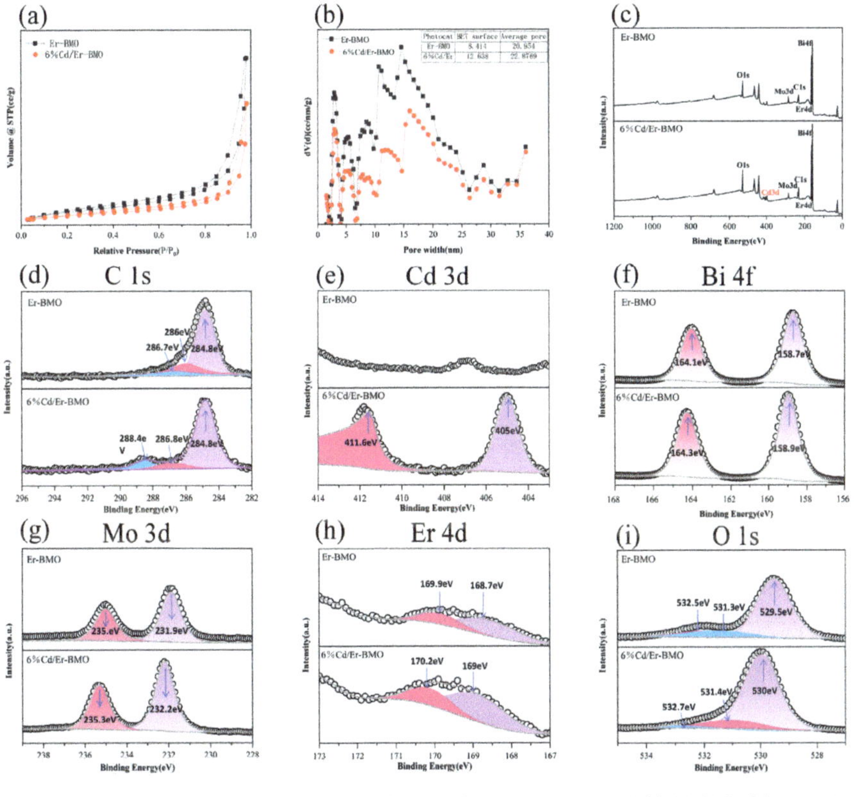

Figure 2. (**a**) N_2 adsorption/desorption isotherms of Er-BMO and 6% Cd/Er-BMO; (**b**) pore size distribution of Er-BMO and 6% Cd/Er-BMO and the XPS spectra of Er-BMO and 6% Cd/Er-BMO; (**c**) full spectra of Er-BMO and 6% Cd/Er-BMO; (**d**) C 1s; (**e**) Cd 3d; (**f**) Bi 4f; (**g**) Mo 3d; (**h**) Er 4d; (**i**) O 1s.

The photocatalytic performance of materials is closely related to the recombination of electrons and holes as the photoexcitation, generation, separation, and recombination of electron-hole pairs have a significant impact on the efficiency of the photocatalytic reaction.

As such, the electron-hole-related properties are analyzed by measuring the UV-visible diffuse reflectance spectrum, transient photocurrent response, and photoluminescence spectrum to understand the optical properties of the material in one step. As shown in Figure 3a, the UV spectrum of Er-BMO has a band gap absorption edge near 471.23 nm. Compared with Er-BMO, upon the increase in the Cd^{2+} doping ratio, the light absorption range of Cd/Er-BMO in the visible light region gradually undergoes a slight red shift, indicating that its light absorption capacity continues to increase and its photocatalytic efficiency continues to improve. The red shift phenomenon stops at 6% Cd/Er-BMO (493.30–513.19 nm). Upon the continued increase in the Cd^{2+} concentration, a slight blue shift (513.19–484.11 nm) gradually occurs. The red shift indicates that Cd^{2+} doping can broaden the light absorption range of Er-BMO and enhance visible light absorption, which leads to the generation of more electrons and holes and, eventually, may improve the photocatalytic reaction efficiency [39]. The subsequent blue shift can be explained by the increased Cd^{2+} doping concentration. An excessive exposure resulted in a change in the Er-BMO structure. In addition, new absorption peaks were found at 481.8 and 614.8 nm, which can be attributed to the transition of erbium ions $4F_{3/2,\,5/2}$ and $4F_{9/2}$ from the $4I_{15/2}$ ground state to the excited state, further confirming that Er^{3+} can provide the energy upconversion centre and promote the transfer of electrons in the system [38]. By plotting the relationship between $(\alpha h\nu)^2$ and photon energy ($h\nu$), the band gap energy (Eg) of the Cd^{2+}-doped Er-BMO photocatalyst can be calculated; this is shown in Figure 3b. In the figure, α, h, and ν are the absorption coefficient, Planck constant, and optical frequency, respectively. As can be seen from Figure 3b, the band gaps of Er-BMO, 2% Cd/Er-BMO, 4% Cd/Er-BMO, 6% Cd/Er-BMO, 8% Cd/Er-BMO, and 10% Cd/Er-BMO are 2.81, 2.72, 2.63, 2.60, 2.74, and 2.78 eV, respectively. The energy band of the Cd^{2+}-doped Er-BMO photocatalyst is reduced in line with the trend in adsorption edge, and the minimum energy band is displayed at 6% Cd/Er-BMO. The smaller the band, the narrower the band width. The narrow band width can improve the photocatalytic efficiency. Therefore, 6% Cd/Er-BMO showed the largest photocatalytic activity [26].

The transient photocurrent method is a widely used characterization technique to detect charge carrier photogeneration and extraction kinetics in optoelectronic devices. In order to illustrate the carrier separation efficiency, the response speed [40] and the calculated transient photocurrent response is shown in Figure 3c. As shown in Figure 3c, the magnitude of the photoresponse intensity was 6% Cd/Er-BMO > 2% Cd/Er-BMO > 10% Cd/Er-BMO > 8% Cd/Er-BMO > 4% Cd/Er-BMO > Er-BMO, among which the photoresponse intensity of 6% Cd/Er-BMO was the largest. This proves that the content of electron-hole pairs generated by its photoexcitation is the highest. Compared with that of Er-BMO, the photoresponse intensity of Cd/Er-BMO increased to varying degrees, indicating that Er-BMO crystals can generate more electron-hole pairs under photoexcitation when doped with Cd^{2+}, thereby further promoting the photocatalytic reaction. PL is a non-destructive and non-contact optical method for probing the electronic structure of materials. When light hits the sample, it begins a process called photoexcitation, in which the light is absorbed and excess energy is given to the material [41]. In addition, photoluminescence (PL) spectroscopy can be used to further analyze the separation and transport of photogenerated charges in materials and to further study the recombination efficiency of photocarriers in photocatalysis [42]. This is the spectrum of the intensity or energy distribution of light with different wavelengths formed by the recombination of electrons and holes in the quasi-equilibrium state of the material. The higher the peak value, the higher the electron-hole recombination efficiency of the material and the worse the photocatalytic performance [43]. The PL spectra of Er-BMO and Cd/Er-BMO are shown in Figure 3d. The typical response peaks are identified in the emission spectra of Er-BMO and Cd/Er-BMO. Compared with the strong response peak of Er-BMO, Cd-BMO shows a relatively weak peak (the intensity of 10% Cd/Er-BMO is the highest), which may be attributable to an excessive doping concentration leading to the photoinduced recombination between electrons and holes and the subsequent increase in the PL intensity [44].

This indicated that the appropriate amount of Cd^{2+} doping can promote the transfer of electrons and holes. In summary, Cd/Er-BMO can effectively promote the generation of photogenerated carriers, which inhibit and improve the recombination and separation of electron-hole pairs by doping Cd ions in Er-BMO. In particular, 6% Cd/Er-BMO showed a significantly improved electron excitation and transfer ability and revealed an excellent photocatalytic degradation performance.

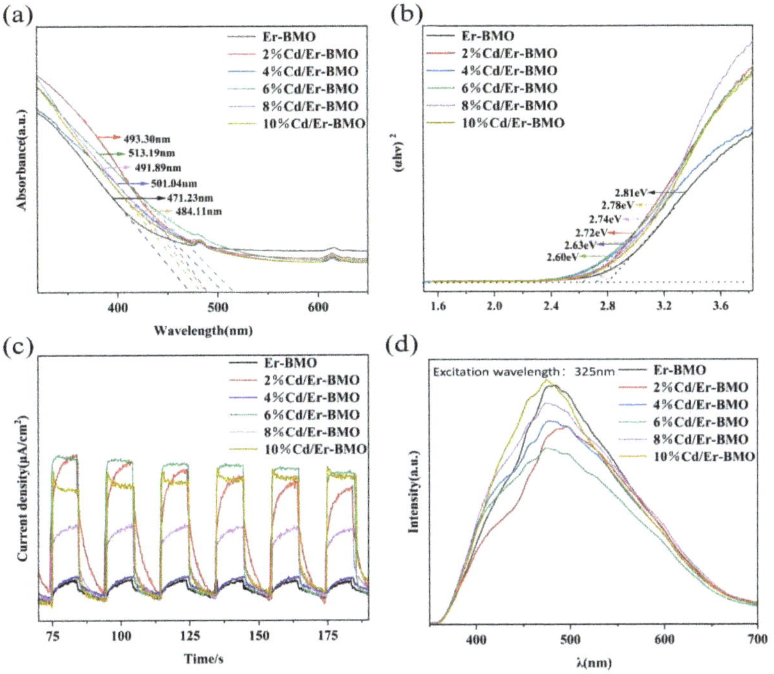

Figure 3. (**a**) UV-Vis DRS absorption spectra of Er-BMO and Cd/Er-BMO; (**b**) $(\alpha h\nu)^2$and hν curves; (**c**) transient photocurrent response of Er-BMO and Cd/Er-BMO; (**d**) photoluminescence spectra of Er-BMO and Cd/Er-BMO.

3.2. Photocatalytic Degradation

In order to explore the photocatalytic pollutant degradation ability of Er-BMO and Cd/Er-BMO with different doping ratios, 5 mg·L^{-1} SMZ was photocatalytically degraded under simulated visible light radiation under a 400 W long arc xenon lamp. Meanwhile, recycling degradation experiments and free radical capture experiments were conducted to study the stability of bismuth molybdate materials and the free radical degradation mechanism. For the repeatability of the experiment, we tested the irradiation intensity of the catalyst in the experiment, which is 4.72×10^4 Lux (Err: 0.13×10^4 Lux).

Firstly, the degradation efficiency diagram of the Er-BMO and Cd/Er-BMO crystals for the SMZ solution under simulated visible light irradiation to study the photocatalytic degradation performance is shown in Figure 4a. The relative concentration ($C_0 - C/C_0$) of the SMZ solution (each point separated by 45 min) was plotted as a function of reaction time. The concentration of pollutant solution was measured using a UV-visible spectrophotometer, and the degradation efficiency under the visible illumination time can be calculated according to the following formula:

$$\eta = \frac{C_0 - C}{C_0} \times 100\%$$

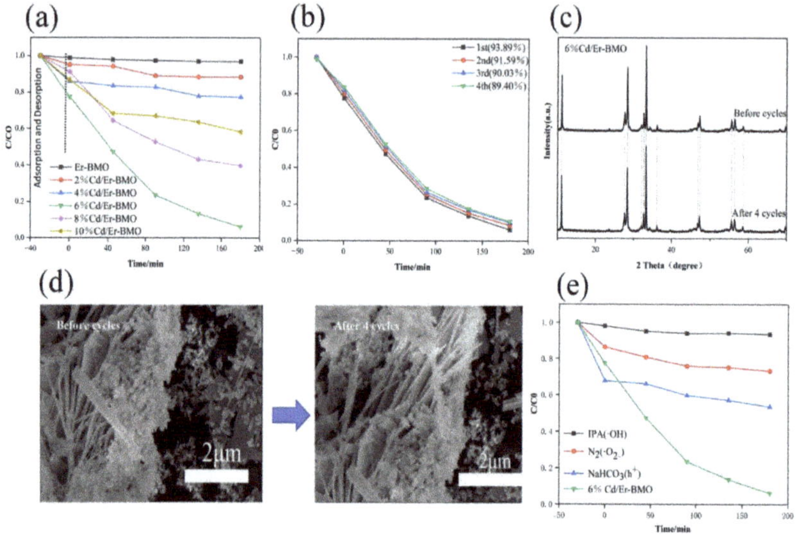

Figure 4. (**a**) Experimental study on photocatalytic degradation of SMZ; (**b**) recycling and degradation experiment of 6% Cd/Er-BMO (SMZ); (**c**) XRD images before and after recycling and degradation experiment; (**d**) SEM images before and after degradation experiment; (**e**) capture experiment of 6% Cd/Er-BMO (SMZ).

As shown in the figure, the adsorption degree of SMZ at 6% Cd/Er-MBO (more than 20%) is large, while the adsorption on Er-MBO can be ignored, which proves that the existence of Cd^{2+} is related to the adsorption capacity. This may be related to surface acidity, as the addition of Cd^{2+} leads to an increase in the number of oxygen vacancies. The degradation efficiency of Er-BMO for SMZ was only 3.11% within the first 180 min. Other than that, the degradation efficiency of SMZ by BMO was only 4.1% in 210 min, which proved that pure BMO was not active in the degradation process [18]. After adding Cd^{2+}, the degradation efficiency was significantly improved with 6% Cd/Er-BMO revealing the most obvious increase, which reached 93.89% degradation efficiency in 180 min. For different contents of Cd/Er-BMO, the order of the degradation efficiencies is 6% Cd/Er-BMO > 8% Cd/Er-BMO > 10% Cd/Er-BMO > 4% Cd/Er-BMO > 2% Cd/Er-BMO. The results fully reflect the good photocatalytic ability of Cd/Er-BMO.

The repeated degradation experiment of 6% Cd/Er-BMO is shown in Figure 4b. Four repeated photocatalytic experiments were carried out on 6% Cd/Er-BMO under the same reaction conditions to investigate the degradation of SMZ. That is, after the experiment was completed, the sample was recovered and the experiment was repeated for a total of four times. After one experiment, the sample is usually rinsed with water and dried before the next experiment. There will be some losses during the experiment. We will reduce the amount of pollutant solution according to the quality of the remaining samples in the previous proportion to compensate for the losses in repeated experiments. The results showed that the degradation efficiencies of the four experiments were 93.89%, 91.59%, 90.03%, and 89.40%, respectively, indicating its good sustainability, durability, and stability. In addition, the 6% Cd/Er-BMO photocatalyst after the degradation experiment was characterized using XRD and SEM (Figure 4c,d). A comparison before and after the experiment showed that the peaks shown in the XRD spectrum of 6% Cd/Er-BMO were still consistent. The SEM image suggested the 6% Cd/Er-BMO maintained its morphology, which confirmed its stability and great potential.

At the same time, ICP-MS was used to detect the concentration of ions in the solution after photocatalytic degradation, and the concentration of the specified element was obtained using the correlation function. The test results and related functions are shown in Table 1. Cd^{2+} (0.0835 mg·L^{-1}) and Bi^{3+} (0.4839 mg·L^{-1}) showed that there was a small amount of ion overflow during photocatalytic degradation, but the photocatalytic efficiency was not affected. It should be noted that the Cd^{2+} concentration shown in the table is 0.0835 mL·L^{-1}, which is lower than the continuous emission concentration of the Chinese government (0.1 mg·L^{-1}) (ref: GB 8978-1996; China National Comprehensive sewage discharge standard). In summary, our experiments were carried out under prescribed and reasonable conditions, which also indicated that 6% Cd/Er-BMO is an excellent photocatalytic material. Nevertheless, the concentration of 0.0835 mg·L^{-1} is higher than the limit for drinking water (0.005 mg·L^{-1}) and irrigation water (0.01 mg·L^{-1}), demonstrating that the solution still needs further treatment before it can be used for drinking and irrigation, and that the presence of Cd^{2+} in the photocatalyst seems to place some limits on the application of this catalyst.

Table 1. ICP-MS experiment.

Sample	Cd^{2+} Intensity	Cd^{2+} Concentration (mg·L^{-1})	Bi^{3+} Intensity	Bi^{3+} Concentration (mg·L^{-1})
1 mg·L^{-1} Cd^{2+} solution	20,414.82	1.0124	\	\
1 mg·L^{-1} Bi^{3+} solution	\	\	569.90	1.0357
6% Cd/Er-BMO degradation solution	1684.75	0.0835	266.90	0.4839

Cd^{2+} Intensity = 20,163.71 × Concentration + 1.081; Bi^{3+} Intensity = 549.02 × Concentration + 1.227.

A capture experiment is an experimental method to detect and study particles, molecules, ions, or reaction intermediates. Free radicals and holes play a key role in many biological and chemical processes. Therefore, studying free radicals and holes plays an important role in understanding these processes. As shown in Figure 4e, the formation and reaction mechanism of common functional groups and holes during the degradation of SMZ solution were studied through capture experiments. Since hydroxyl radicals (·OH), superoxide radicals (·O_{2-}), and holes (h^+) are the most active in photocatalysis, 1 mM isopropanol (IPA), 1 mM sodium bicarbonate ($NaHCO_3$), and nitrogen (N_2) are used as capture agents to capture ·OH, h^+, and ·O_{2-} functional groups. In addition, nitrogen was continuously introduced through a tube during the experiment to capture superoxide radicals. When IPA, $NaHCO_3$, and N_2 were added to the experiment, the photocatalytic degradation performance decreased significantly. Compared with the reaction system without the addition of a capture agent, the degradation effect of 6% Cd/Er-BMO decreased from 93.89% to 6.64% after adding IPA as a capture agent during the degradation process, which seriously affected the photocatalytic reaction activity. Secondly, the degradation effect dropped from 93.89% to 26.79% after adding nitrogen. After adding $NaHCO_3$, the degradation effect dropped from 93.89% to 46.43%. The results indicated that after co-doping, the ·OH, h^+, and ·O_{2-} radicals are still important active substances in the photocatalytic degradation of organic matter, and they play an important role in the degradation of SMZ. According to previous studies, the addition of IPA in the case of doping with only Cd^{2+} did not significantly change the photocatalytic degradation performance [18], which further confirmed that the doping of Er^{3+} benefits the formation of ·OH [23].

3.3. Photocatalytic Mechanism Analysis

The photocatalytic mechanism refers to the process and principle of photocatalysts inducing chemical reactions under light conditions, the study of which is conducive to understanding the nature of the photocatalytic process, optimizing the design of catalysts, and improving the efficiency of photocatalysis. Semiconductor photoexcitation as a photocatalytic mechanism refers to the band structure of semiconductor particles, which consists of a low-energy valence band (VB) filled with electrons and an empty high-energy conduction band (CB) with a forbidden band between them. When a light source of appropriate energy irradiates the semiconductor, the semiconductor will be activated as a result of the absorption of the energy of the photon, resulting in electron jumping from the valence band to the conduction band, generating electron-hole pairs. The electron-hole pairs separate and migrate to the surface of the semiconductor, and some of them recombine during the migration process, losing their activity. As such, to improve the quantum yield of semiconductor photocatalytic reactions, it is necessary to make the acceptor potential lower than the semiconductor conduction band potential, as well as making the donor potential higher than the semiconductor valence band potential, to effectively inhibit the direct recombination of photogenerated electrons and holes.

For the co-doping system in this paper, when the sample was irradiated with light with energy equal to or greater than the band gap energy, e^- and h^+ are formed and exist in the conduction band and valence band, respectively. As Cd^{2+} was doped in Er-BMO, especially 6% Cd/Er-BMO, the band gap was decreased, the generation of electron-hole pairs was promoted with the direct recombination of photogenerated electrons, and the holes were effectively suppressed. As such, a large number of e^- and h^+ were separated to the appropriate position to contact the pollutants, improving the degradation efficiency. $\cdot O_{2-}$ was developed when e^- contacts O_2, and then the h^+ and active radical ($\cdot O_{2-}$ and $\cdot OH$) were employed for the SMZ degradation based on the free radical capture experiment results. Note that h^+ is mainly used directly for the degradation experiments, as opposed to using $\cdot OH$ formed by h^+. Thus, the possible degradation process is as follows: SMZ is hydrolyzed during the reaction to form the product TP 1, then h^+ and $\cdot OH$ break the S-N bond to form the products TP 2 and TP 3. Finally, all intermediate products further react with h^+ and $\cdot O_{2-}$ to form smaller molecules or mineralize to CO_2 and H_2O [2]. In addition, in order to simplify the picture, we will replace the pollutant in the left half of Figure 5 with R-H, which will generate H_2O and simple organic matter (R·) during the reaction process. The detailed photodegradation process is as follows:

$$\mathrm{Cd/Er-BMO} + h\mu \rightarrow \mathrm{Cd/Er-BMO}(e^{-}cb + h^{+}vb) \quad (1)$$

$$e^{-}cb + O_2 \rightarrow \cdot O_{2-} \quad (2)$$

$$\cdot O_{2-} + H_2O \rightarrow \cdot OH; h^{+}vb + H_2O \rightarrow \cdot OH \quad (3)$$

$$\begin{gathered} \mathrm{SMZ} + h^{+}vb + \cdot OH \rightarrow C_9H_8N_2O_3S \\ C_9H_8N_2O_3S + h^{+}vb + \cdot OH \rightarrow C_3H_4N_2O + C_6H_6O_2S \\ C_3H_4N_2O + h^{+}/\cdot O_{2-} \rightarrow H_2O + CO_2 \\ C_6H_6O_2S + h^{+}/\cdot O_{2-} \rightarrow H_2O + CO_2 \end{gathered} \quad (4)$$

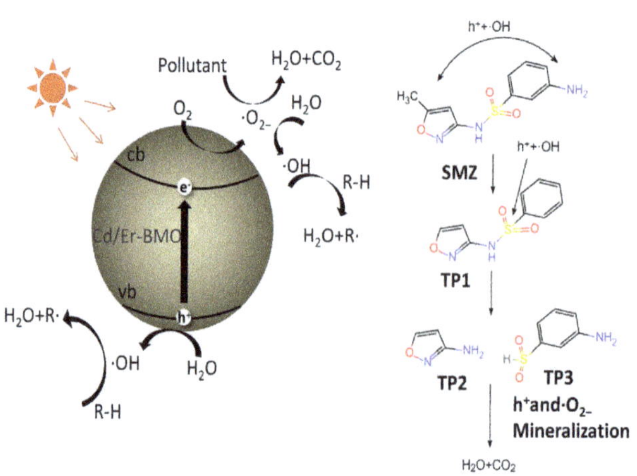

Figure 5. The mechanism of the photocatalytic degradation of Cd/Er-BMO.

4. Conclusions

In this paper, a simple hydrothermal method was used to successfully prepare the photocatalyst Cd/Er-BMO, which can be applied to the photocatalytic degradation of antibiotics and can improve the efficiency of pollutant degradation. Compared with Er-BMO, Cd/Er-BMO has a higher specific surface area, improved electronic properties, and a higher catalytic activity. In addition, Cd/Er-BMO did not produce any other crystal phases during the doping process. Through the analysis of the UV-visible diffuse reflectance spectrum, transient photocurrent response, and photoluminescence spectrum, it is confirmed that 6% Cd/Er-BMO can not only effectively promote the generation of photogenerated carriers, but also inhibit the recombination of electron-hole pairs and their separation, thereby further enhancing the photocatalytic degradation performance of Cd/Er-BMO. In repeated degradation experiments, 6% Cd/Er-BMO showed an excellent sustainability and stability, confirming that the material can maintain a good degradation ability during photocatalytic degradation. Free radical capture experiments revealed that h^+, $\cdot O_2-$, and $\cdot OH$ were the main active substances for the effective degradation of SMZ, further confirming the effectiveness and advantages of Cd/Er-BMO in practical applications. In summary, this study systematically investigated the degradation process of SMZ doped with Cd^{2+} Er-BMO, and found that 6% Cd/Er-BMO demonstrated an excellent catalytic performance. This work identified a new effective photocatalyst for real-world degradation and provided ideas and directions for the future research and development of more efficient photocatalysts.

Author Contributions: Conceptualization, N.Y. and F.Z.; formal analysis, Y.N.; methodology, B.Z.; project administration, B.Z.; resources, N.Y. and F.Z.; visualization, B.Z. and Y.N.; writing—original draft, B.Z. and Y.N.; writing—review and editing, N.Y. and F.Z. All authors have read and agreed to the published version of the manuscript.

Funding: This research was funded by the National Natural Science Foundation of China, grant number No. 62264015.

Institutional Review Board Statement: Not applicable.

Informed Consent Statement: Not applicable.

Data Availability Statement: Data are contained within this article.

Acknowledgments: We acknowledge the Network Information Center of Yan'an University for providing supercomputing resources and related support for this study.

Conflicts of Interest: The authors declare no conflicts of interest.

References

1. Dai, J.J.; Song, J.M. Environmental biogeochemical characteristics of sulfonamides in typical aquatic environments of China. *Oceanol. Et Limnol. Sin.* **2023**, *54*, 935–950.
2. Wang, A.; Ni, J. MOF-derived N-doped ZnO carbon skeleton@hierarchical Bi_2MoO_6 S-scheme heterojunction for photodegradation of SMX: Mechanism, pathways and DFT calculation. *J. Hazard. Mater.* **2022**, *426*, 128106. [CrossRef] [PubMed]
3. Qiu, T.; Chen, G. Degradation of naphthalene and anthracene using the CdS/ZnO with enhanced photocatalytic activities. *Environ. Technol. Innov.* **2024**, *36*, 103741. [CrossRef]
4. Mao, T.; Zha, J. Research Progress of TiO_2 Modification and Photodegradation of Organic Pollutants. *Inorganics* **2024**, *12*, 178. [CrossRef]
5. Waldner, G.; Pourmodjib, M. Photoelectrocatalytic degradation of 4-chlorophenol and oxalic acid on titanium dioxide electrodes. *Chemosphere* **2003**, *50*, 989–998. [CrossRef]
6. Kshirsagar, A.S.; Khanna, P.K. Titanium dioxide (TiO_2)-decorated silver indium diselenide ($AgInSe_2$): Novel Nano-Photocatalyst for Oxidative Dye Degradation. *Inorg. Chem. Front.* **2018**, *5*, 2242–2256. [CrossRef]
7. Brahimi, R.; Bessekhouad, Y. Visible light induced hydrogen evolution over the heterosystem Bi_2S_3/TiO_2. *Catal. Today* **2007**, *122*, 62–65. [CrossRef]
8. Bai, Y.; Chen, H. Coupling physical adsorption and photocatalysis over CdS/UiO-66-NH_2 for efficient removal of hydrogen sulfide. *Sep. Purif. Technol.* **2024**, *341*, 126956. [CrossRef]
9. Sun, X.; He, K. Construction of visible-light-response photocatalysis-self-Fenton system for the efficient degradation of amoxicillin based on industrial waste red mud/CdS S-scheme heterojunction. *Sep. Purif. Technol.* **2023**, *324*, 124600. [CrossRef]

10. Gopal, V.; Harsha, S. Cow urine-based green synthesis of sunlight-responsive ZrO_2-Bi_2O_3 and its application in photocatalysis of 2,4-Dichlorophenoxyacetic acid in aqueous solution—Kinetics, mechanisms and sustainability analysis. *Catal. Commun.* **2024**, *187*, 106869. [CrossRef]
11. Wang, B.; Guo, Y. Design of Porous ZrO_2 with Well-Tuned Band Structures and Strong Visible-Light Harvesting via Zn Doping for Enhanced Visible-Light Photocatalysis. *Chem. Eng. J.* **2024**, *481*, 148489. [CrossRef]
12. Perciani de Moraes, N.; Amarante Pereira, R. Cross-linked cellulose beads as an eco-friendly support for ZnO/SnO_2/carbon xerogel hybrid photocatalyst: Exploring the synergy between adsorption and photocatalysis under simulated sunlight. *Int. J. Biol. Macromol.* **2024**, *254*, 127826.
13. Mahanta, R.; Chetri, P. Investigation of visible light photocatalysis effect of SnO_2 nanoparticles. *Mater. Today Proc.* **2023**, *1*, 348. [CrossRef]
14. Hamza, M.A.; El-Shazly, A.N. Novel Bi-based photocatalysts with unprecedented visible light-driven hydrogen production rate: Experimental and DFT insights. *Chem. Eng. J.* **2020**, *384*, 123351. [CrossRef]
15. Tsay, C.-Y.; Chung, C.-Y. Fe-Doped g-C_3N_4/Bi_2MoO_6 Heterostructured Composition with Improved Visible Photocatalytic Activity for Rhodamine B Degradation. *Molecules* **2024**, *29*, 2631. [CrossRef]
16. Zhang, B.; Liu, G. Density Functional Theory Study of Electronic Structure and Optical Properties of Ln^{3+}-Doped γ-Bi_2MoO_6 (Ln=Gd, Ho, Yb). *Crystals* **2023**, *13*, 1158. [CrossRef]
17. Ning, J.; Zhang, B.H. Designing Advanced S-Scheme CdS QDs/La-Bi_2WO_6 Photocatalysts for Efficient Degradation of RhB. *Exploration* **2023**, *3*, 20230050. [CrossRef]
18. Zhang, B.; Fang, C. Unusual aliovalent Cd doped γ-Bi_2MoO_6 nanomaterial for efficient photocatalytic degradation of sulfamethoxazole and rhodamine B under visible light irradiation. *Carbon Neutralization* **2023**, *2*, 646–660. [CrossRef]
19. Wang, J.F.; Zhao, C.R. One-step fabrication of Cu-doped Bi_2MoO_6 microflower for enhancing performance in photocatalytic nitrogen fixation. *J. Colloid. Interface Sci.* **2023**, *638*, 427–438. [CrossRef]
20. Meng, Q.Q.; Lv, C. High-efficiency Fe-Mediated Bi_2MoO_6 nitrogen-fixing photocatalyst: Reduced surface work function and ameliorated surface reaction. *Appl. Catal. B Environ.* **2019**, *256*, 117781. [CrossRef]
21. Li, A.; Li, H. Structure, properties and application of Eu^{3+} doped bismuth molybdate red-emitting phosphor synthesized by sol-gel method. *Opt. Mater.* **2023**, *144*, 114336. [CrossRef]
22. Xu, J.; Liu, Y. Preparation of Cd-Doped Bi_2MoO_6 Photocatalyst for Efficient Degradation of Ofloxacin under Visible Light Irradiation. *Surf. Interfaces* **2021**, *25*, 101246. [CrossRef]
23. Li, H.D.; Li, W.J. Engineering of Gd/Er/Lu-triple-doped Bi_2MoO_6 to synergistically boost the photocatalytic performance in three different aspects: Oxidizability, light absorption and charge separation. *Appl. Surf. Sci.* **2019**, *463*, 556–565. [CrossRef]
24. Wang, M.; Guo, P. The honeycomb-like Eu^{3+}, Fe^{3+} doping bismuth molybdate photocatalyst with enhanced performance prepared by a citric acid complex process. *Mater. Lett.* **2017**, *192*, 96–100. [CrossRef]
25. Li, S.; Hasan, N. 2D Bi_2MoO_6/$Zn_3V_2O_8$ heterojunction photocatalyst for efficient photocatalytic reduction of CO_2 to CO and CH_4. *J. Colloid. Interface Sci.* **2023**, *652*, 1533–1544. [CrossRef]
26. Sajjadi, S.M.; Hossinzadeh, G. Textile dyes removing from the wastewater by green synthesized Cu-doped ZnO photocatalysts under the simulated sunlight illumination. *Ceram. Int.* **2024**, *7*, 11. [CrossRef]
27. Chen, S.; Li, Y. Enhanced Photocatalytic Activity of Te-Doped Bi_2MoO_6 under Visible Light Irradiation: Effective Separation of Photogenerated Carriers Resulted from Inhomogeneous Lattice Distortion and Improved Electron Capturing Ability. *J. Solid. State Chem.* **2017**, *249*, 124–130. [CrossRef]
28. Xing, Y.; Gao, X. Synthesis of carbon doped Bi_2MoO_6 for enhanced photocatalytic performance and tumor photodynamic therapy efficiency. *Appl. Surf. Sci.* **2019**, *465*, 369–382. [CrossRef]
29. Barsotti, E.; Tan, S.P. A review on capillary condensation in nanoporous media: Implications for hydrocarbon recovery from tight reservoirs. *Fuel* **2016**, *184*, 344–361. [CrossRef]
30. Liu, X.; Zheng, L. Identifying the Activity Origin of a Cobalt Single-Atom Catalyst for Hydrogen Evolution Using Supervised Learning. *Adv. Funct. Mater.* **2021**, *31*, 10. [CrossRef]
31. Phasayavan, W.; Japa, M. Oxygen-Deficient Bismuth Molybdate Nanocatalysts: Synergistic Effects in Boosting Photocatalytic Oxidative Coupling of Benzylamine and Mechanistic Insight. *J. Colloid. Interface Sci.* **2021**, *581*, 719–728. [CrossRef] [PubMed]
32. Liu, X.; Xi, S. Restructuring highly electron-deficient metal-metal oxides for boosting stability in acidic oxygen evolution reaction. *Nat. Commun.* **2021**, *12*, 5676. [CrossRef] [PubMed]
33. Guo, F.; Chen, Z. Ternary Ni_2P/Bi_2MoO_6/g-C_3N_4 composite with Z-scheme electron transfer path for enhanced removal broad-spectrum antibiotics by the synergistic effect of adsorption and photocatalysis. *Chin. J. Chem. Eng.* **2021**, *44*, 157–168. [CrossRef]
34. Xie, R.; Fan, J. Hierarchical Bi_2MoO_6 microsphere photocatalysts modified with polypyrrole conjugated polymer for efficient decontamination of organic pollutants. *Chemosphere* **2022**, *286*, 131541. [CrossRef] [PubMed]
35. Frankcombe, T.; Liu, Y. Interpretation of Oxygen 1s X-ray Photoelectron Spectroscopy of ZnO. *Materials* **2023**, *16*, 1234. [CrossRef]
36. Wang, G.; Huo, T. Surface-layer bromine doping enhanced generation of surface oxygen vacancies in bismuth molybdate for efficient photocatalytic nitrogen fixation. *Appl. Catal. B Environ.* **2022**, *310*, 121319. [CrossRef]
37. Liu, W.; Yin, K. In situ synthesis of Bi_2MoO_6@C@attapulgite photocatalyst for enhanced photocatalytic nitrogen fixation ability under simulated solar irradiation. *Colloids Surf. A Physicochem. Eng. Asp.* **2020**, *591*, 124488. [CrossRef]

38. Reszczyńska, J.; Grzyb, T. Visible light activity of rare earth metal doped (Er^{3+}, Yb^{3+} or Er^{3+}/Yb^{3+}) titania photocatalysts. *Appl. Catal. B Environ.* **2015**, *163*, 40–49. [CrossRef]
39. Zhang, Y.; Wang, M. Ternary heterojunction of cross-linked benzene polymer/Bi_2MoO_6-graphene oxide catalysts promote efficient adsorption and photocatalytic removal of oxytetracycline. *J. Colloid. Interface Sci.* **2024**, *668*, 437–447. [CrossRef]
40. Sun, Z.; Yang, X. Surface oxygen vacancies of Pd/Bi_2MoO_{6-x} acts as "Electron Bridge" to promote photocatalytic selective oxidation of alcohol. *Appl. Catal. B Environ.* **2021**, *285*, 119790. [CrossRef]
41. Mitrić, J. Properties and characterization of rare-earth-activated phosphors. In *Rare-Earth-Activated Phosphors*; Elsevier: Amsterdam, The Netherlands, 2022; pp. 43–58.
42. Ke, J.; Duan, X. UV-assisted construction of 3D hierarchical rGO/Bi_2MoO_6 composites for enhanced photocatalytic water oxidation. *Chem. Eng. J.* **2017**, *313*, 1447–1453. [CrossRef]
43. Geng, L.; Li, W. Active sites modification and superior carriers separation synergistically boosted hydrogen production of Bi/Bi_2MoO_6/$ZnIn_2S_4$ non-noble metal S-scheme photocatalyst. *J. Colloid. Interface Sci.* **2023**, *629*, 723–732. [CrossRef] [PubMed]
44. Ahmadi, M.; Alavi, S.M. Effective CO_2 photoreduction to methane over Bi_2MoO_6/Ni, N co-doped TiO_2 nano-photocatalyst. *Int. J. Hydrogen Energy* **2024**, *56*, 1309–1323. [CrossRef]

coatings

MDPI

Article

Metal Ions' Dynamic Effect on Metal-Assisted Catalyzed Etching of Silicon in Acid Solution

Xiaoyu Yang †, Ying Liu †, Lin Wu, Zhiyuan Liao, Baoguo Zhang, Tinashe Tembo, Yichen Wang and Ya Hu *

Key Laboratory of Hubei Province for Coal Conversion and New Carbon Materials, School of Chemistry and Chemical Engineering, Wuhan University of Science and Technology, Wuhan 430081, China; zhlyxy@wust.edu.cn (X.Y.); liuying-9325@wust.edu.cn (Y.L.); wulin@wust.edu.cn (L.W.); liaozhiyuan@wust.edu.cn (Z.L.); zbg@wust.edu.cn (B.Z.); tinashephextembo1165@gmail.com (T.T.); yichenwang2413@163.com (Y.W.)

* Correspondence: huya@wust.edu.cn

† These authors contributed equally to this work.

Abstract: Metal-assisted catalyzed etching (MACE) technology is convenient and efficient for fabricating large-area silicon nanowires at room temperature. However, the mechanism requires further exploration, particularly the dynamic effect of various ions in the acid-etching solution. This paper investigated the MACE of silicon wafers predeposited with metal nanofilms in an HF-M(NO_3)x-H_2O etching solution (where M(NO_3)x is the nitrate of the fourth-period elements of the periodic table). The oxidizing ability of Fe^{3+} and NO_3^- was demonstrated, and the dynamic influence of metal ions on the etching process was discussed. The results show that the MACE of silicon can be realized in various HF-M$(NO_3)_x$-H_2O etching solutions, such as KNO_3, $Al(NO_3)_3$, $Cr(NO_3)_3$, $Mn(NO_3)_2$, $Ni(NO_3)_2$, $Co(NO_3)_2$, HNO_3, and $Ca(NO_3)_2$. It is confirmed that the concentration and type of cations in the etching solution affect the etching rate and morphology of silicon. Fe^{3+} and NO_3^- act as oxidants in catalytic etching. The fastest etching rate is about 5~6 μm/h in $Ni(NO_3)_2$, $Co(NO_3)_2$, and $Ca(NO_3)_2$ etching solutions. However, a high concentration of K^+ hinders silicon etching. This study expands the application of MACE etching solution systems.

Keywords: dynamic effect; MACE; metal ions; acid solution; nitrate

1. Introduction

Silicon nanowires (SiNWs) have attracted widespread attention in the scientific community in recent decades because of their special physical and chemical properties [1–6]. There are various methods for synthesizing SiNWs. In 1996, hydrothermal technology was first used to prepare porous silicon [7,8]. The reactor was filled with the HF-$Fe(NO_3)_3$-H_2O solution and heated by a thermostat to provide high temperature and high pressure to promote the reaction. This hydrothermal technology enables the preparation of porous silicon without an external electric power supply. However, this method requires closed containers like autoclaves. In 2005, K.Q. Peng deposited a Ag or Au nanoparticle layer by electroless deposition on the surface of the silicon wafer, immersed the processed silicon wafer in an HF-$Fe(NO_3)_3$-H_2O solution at room temperature for half an hour, and successfully synthesized large-area SiNWs using metal-assisted catalyzed etching (MACE), which is more orderly [9]. In addition, Peng studied the mechanism of MACE and explained the flow of electrons and holes in the oxidation–reduction etching process of silicon driven by galvanic cells based on band bending theory [9–11]. The contact interface between noble metal particles and silicon can be considered a typical metal–semiconductor contact.

The technology generally adopts a two-step method to prepare nanosilicon by MACE in a liquid system. Firstly, the surface of the cleaned silicon wafer is coated with a noble Ag or Au nanoparticle layer [12–19]. Secondly, the noble-metal-coated silicon wafer is immersed in an etching solution containing HF and an oxidant. The currently reported etching solution systems include HF-H_2O_2-H_2O solution and HF-$Fe(NO_3)_3$-H_2O solution. It is believed that the above-mentioned electroless deposition of the metal nanoparticle layer and MACE of silicon are both localized microelectrochemical redox processes, including the cathode and anode reactions [20]. Taking the electroless deposition of the silver nanoparticle layer on the silicon wafer surface as an example, the first step of MACE in the liquid system, when the clean silicon wafer is immersed in HF-$AgNO_3$-H_2O solution, Ag^+ is oxidation, which injects holes into adjacent silicon to gain electrons, which are reduced to Ag that gradually collects on the surface of the silicon wafer to form a uniform nanoparticle layer. The silicon injected with holes is oxidized and highly unstable in hydrofluoric acid solution and reacts with hydrofluoric acid to form the soluble H_2SiF_6 [21–24]. As the reaction goes on, the silver particles continually grow and accumulate on the surface of the silicon wafer. Finally, the silicon wafer in the plating solution is tightly covered by a uniform film consisting of silver nanoparticles on the surface. The second step of MACE in the liquid system is essentially an electrochemical etching process driven by the galvanic cell. The dispersed Ag nanoparticles deposited on the surface of the silicon wafer combine with the silicon substrate to form a galvanic cell. The galvanic cell can accelerate the etching of the silicon substrate covered by the metal particles, while the uncovered part of the substrate would be retained. Thereby, a neat array of silicon nanowires is formed on the surface. Taking the reaction in the HF-$Fe(NO_3)_3$-H_2O solution system as an example, with the catalysis of silver nanoparticles, the Fe^{3+} can inject a hole into the silicon, obtain electrons, and reduce to Fe^{2+}. As with the reaction in the deposition of the silver nanoparticle layer, holes are injected into the silicon atoms covered by the silver nanoparticles, and the silicon atoms are oxidized. Therefore, the location of the silicon atom at the bottom of the silver nanoparticle becomes a small hole into which the silver nanoparticle "sinks". As the reaction progresses, more Fe^{3+} ions in the solution inject holes into silicon atoms through the silver particles. The silicon atoms are continually oxidized to SiO_2, which reacts with HF to form silicofluoride radical dissolved in solution. The pits at the bottom of the silver nanoparticles continue to deepen, and the silver nanoparticles gradually sink into the interior of the silicon substrate as the pits grow.

However, it is worth noting that the noble Ag nanoparticle layers are not as stable as they are supposed to be, especially when the oxidant is strong, as there is a dynamic equilibrium of deposition and dissolution of the noble metal particle films in the etching solution. Furthermore, various ions lead to a complex reaction equilibrium, as the Fe^{3+} and the NO_3^- ions in the etching solution are also oxidizing but neglected. Focusing on these questions, the objective of this work is to verify the role of Fe^{3+} as the oxidant in the MACE process and to further discuss whether NO_3^- can inject holes into the silicon substrate as the oxidant in MACE. This work also focuses on the influence of the dynamic effect of metal ions in $M(NO_3)_x$ on the MACE process to further explore the mechanism of the MACE process, which helps the expansion of the application and building of low-cost and easy-to-operate systems to prepare large-area SiNW arrays.

2. Materials and Methods

One-side-polished single crystalline N-type (100) 2–2.7 Ω·cm silicon wafers were selected and purchased from Beijing General Research Institute for Nonferrous Metals (GRINM). CH_3COCH_3, C_2H_5OH, and HF were purchased from Beijing Sinopharm Chemical Reagent Co., Ltd., Beijing, China, and deionized water was prepared. The other chemicals used were purchased from Sigma–Aldrich (St. Louis, MO, USA) and used without further treatment. The silicon wafers were cut into 2×2 cm^2 and cleaned ultrasonically in deionized (DI) water, CH_3COCH_3, and C_2H_5OH for 10 min, respectively. Then, the silicon wafers were rinsed three times with DI water and immersed in H_2SO_4-H_2O_2 solution at 80 °C for 30 min. Finally, enough DI water was used to clean the remaining acid on the silicon wafers. The silicon wafers were placed in HF-$AgNO_3$-H_2O or HF-$AuCl_4H$-H_2O solution for 2 min in a fume cupboard. Then, the silicon wafer was removed and cleaned with DI water, and then transferred into sealed Teflon-lined autoclaves at a temperature of

50 °C for one hour. The total volume of the etching solution used in all the experiments in this paper was 50 mL. The concentration of HF in the etching solution was 4.6 M if not specially stated in this paper. After the reaction, the silicon wafers were taken out, cleaned, and dried with the SiNWs on the surface. The silicon wafers were cut into 1 × 1 cm^2, and the cross- and top-sectional morphologies were characterized by a high-resolution field-emission scanning electron microscope (SEM, HITACHI, S-4800, Hitachi of Japan, Tokyo, Japan). SiNWs were scraped off with a razor blade, dispersed ultrasonically in ethanol for 10 min, and then dropped on copper mesh to dry before being characterized by transmission electron microscopy (TEM, JEOL JEM-2100, Japan Electronics, Tokyo, Japan). The crystal composition was investigated by X-ray diffraction (XRD, Cu kα, SmartLab SE, Rigaku, Corporation, Tokyo, Japan).

3. Results and Discussion

3.1. Oxidation Validation of Fe^{3+} and NO_3^-

Because of the catalysis of Ag (or Au and Pt), neat SiNW array structures were successfully prepared on both P-type and N-type silicon substrates, according to previous research [25–28]. However, the etching morphology of the P-type and N-type silicon wafers with normal doping levels in the HF-$Fe(NO_3)_3$-H_2O are almost the same, and the crystal orientations of the SiNW arrays are determined by the crystal orientation of the substrates. Therefore, an N-type (100) silicon wafer with a resistivity of 2.2 Ω·cm was used as the research object, with Ag as the catalyst. First, the cleaned silicon wafers without silver plating were directly immersed in an etching solution consisting of 4.6 M HF and 1.5 M $FeCl_3$ (HF-$FeCl_3$-H_2O) and reacted at 50 °C for 30 min. As shown in Figure 1a, a porous layer about 200 nm thick appeared on the surface of the silicon wafer. This indicates that a redox reaction occurred, and a small amount of silicon was dissolved to form a porous layer, which is consistent with the research results of porous silicon. Next, a silver nanoparticle layer was coated on the surface of the cleaned silicon wafer by the electroless deposition technique described above, and the sample was then put into the HF-$FeCl_3$-H_2O solution at 50 °C for 30 min. As shown in Figure 1b,c, there was no sign of etching on the silicon wafer surface, but some micro- and nanospheres appeared. The surfaces of the spheres were not smooth and seemed to have some attachments.

According to relevant reports, these spheres may be AgCl. They were formed from silver nanoparticle layers that dissolved into Ag^+ and combined with Cl^- in the system to form AgCl spheres [29–32]. Then, the sample was soaked in concentrated nitric acid for half an hour. As shown in Figure 1d,e, these small balls still existed and became smoother. It proves that these balls were indeed AgCl balls, and the surface attachments may be silver dissolved in concentrated nitric acid. For further verification, the crystal composition of the samples was characterized by XRD. The clean silicon wafer coated with a silver nanoparticle layer was directly tested. The XRD curve is shown in Figure 1f(a), and only the characteristic peaks of silver appear around 38° and 44.5°. After the silver-plated silicon wafer was immersed in the HF-$FeCl_3$-H_2O solution at 50 °C for 30 min, the characteristic peaks of silver (JCPDS#: 41-1402; the grazing incidence angle was 5°) became weak (Figure 1f(b)), indicating that most silver particles had disappeared.

Compared with the standard XRD pattern, several new characteristic peaks corresponded exactly to those of AgCl (JCPDS#: 31-1238; the grazing incidence angle was 5°), indicating that the small balls in the above pictures were AgCl. Although the silver-plated silicon wafers did not become SiNW arrays when immersed into the HF- $FeCl_3$-H_2O solution, as was initially expected, this does not prove the inability of Fe^{3+} to oxidize. It was suspected that the primary reason for the failure of the expected effect was that silver is unstable in the environment of Fe^{3+} and becomes Ag^+ combined with Cl^-. The results may be different when changing the silver-coated silicon wafer for the gold-coated silicon wafer.

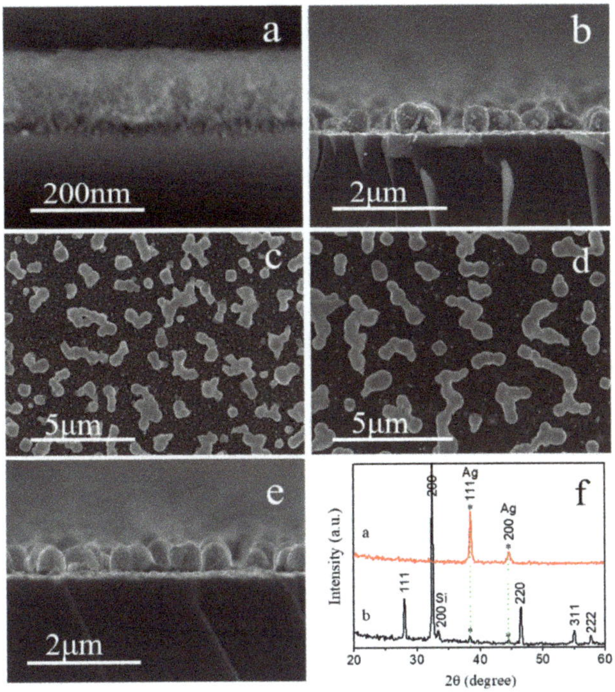

Figure 1. SEM morphology of silicon wafer etched in etching solution of 1.5 M $FeCl_3$ and 4.6 M HF at 50 °C for 30 min: (**a**) silicon etched without silver plating, (**b**) silicon etched with silver plating, (**c**) top view of Figure 1b, (**d**) sample etched with silver plating and immersed in concentrated HNO_3 for 30 min, (**e**) cross-sectional of Figure 1d, (**f**) The XRD curve of the silver-coated silicon surface (red) and silver-coated silicon wafer after reaction in $FeCl_3$-HF solution (black), * represents silver particles.

To verify that Fe^{3+} ions can act as oxidants to etch silicon using electroless deposition technology, a clean N-type silicon wafer (100) with 2–2.7 Ω·cm resistivity was immersed in a plating solution mixed with $AuCl_4H$ and HF for 1 min to plate a gold nanoparticle layer on the surface. Then, the gold-coated silicon wafer was placed in the etching solution mentioned above, consisting of 4.6 M HF and $FeCl_3$ etching solution with concentrations of 0 M, 1 M, 1.5 M, and 2 M, respectively, and reacted at 50 °C for 30 min. As shown in Figure 2a–d, the one-dimensional SiNW array structures were successfully prepared, and with the increase of the concentration of $FeCl_3$, SiNW array structures with different lengths were obtained. The above results indicate that Fe^{3+} is more electronegative than silicon and acts as an oxidant in the HF-$FeCl_3$-H_2O system. When clean silicon wafers are immersed in HF-$AuCl_4H$-H_2O solution, Au^{3+} is oxidizing and able to inject holes into the adjacent silicon to gain electrons and reduce to Au, which gradually gathers on the surface of silicon wafers to form uniform nanoparticle layers. The silicon injected with holes is oxidized and is extremely unstable in HF solution, reacting with HF to form soluble H_2SiF_6 [21–24] and resulting in the structure of SiNWs (Figure 2a). Then, the gold-plated silicon wafer was immersed in the HF-$FeCl_3$-H_2O system. With gold particles as the reactive center, the Fe^{3+} ions in the solution injected holes into the silicon atoms under the gold nanoparticles, extracted electrons, and were reduced to Fe^{2+} ions. The silicon under the gold nanoparticles was oxidized and quickly dissolved in HF, leaving small pits. With the continuous progress of the etching, the small pits kept getting deeper, and the gold nanoparticles also moved to the inside of the silicon substrate along with the pits, leaving the uncoated silicon wafer as a silicon wire array structure.

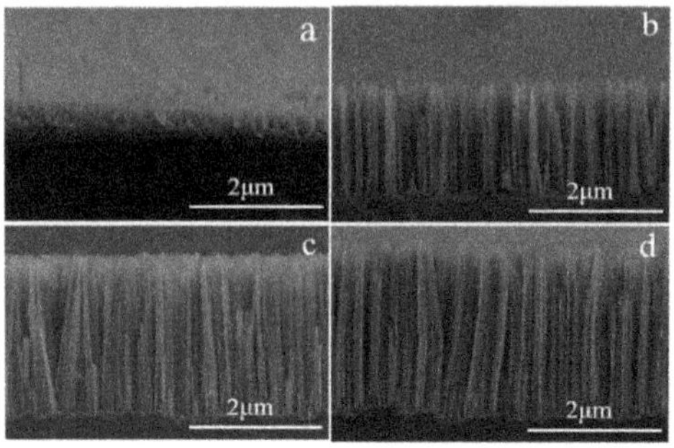

Figure 2. The cross-sectional view of gold-plated silicon wafer etched in 4.6 M HF and different concentrations of $FeCl_3$ etching solution at 50 °C for 30 min: (**a**) 4.6 M HF and 0 M $FeCl_3$, (**b**) 4.6 M HF and 1 M $FeCl_3$, (**c**) 4.6 M HF and 1.5 M $FeCl_3$, and (**d**) 4.6 M HF and 2 M $FeCl_3$.

The above results fully verify that Fe^{3+} can work as the oxidant in the MACE process [9]. To verify the effect of NO_3^-, the silicon wafer with a deposited silver nanoparticle layer was prepared in a solution containing 0.01 M $AgNO_3$ and 4.6 M HF, and the HF-$Fe(NO_3)_3$-H_2O etching solution system was replaced with HF-HNO_3-H_2O, keeping the concentration of NO_3^- consistent. The silicon wafer with silver coating was immersed in two sealed Teflon-lined autoclaves containing HF-$Fe(NO_3)_3$-H_2O and HF-HNO_3-H_2O solution, respectively, with a treatment temperature of 50 °C for one hour. The cross-sectional SEM mages of SiNW etching in HF-$Fe(NO_3)_3$-H_2O and HF-HNO_3-H_2O solution for 1 h are shown in Figures 3a and 3b, respectively. Figure 3a shows that the silicon wafer was exactly etched in the HF-HNO_3-H_2O system. The etching rate of Si in the HF-$Fe(NO_3)_3$-H_2O solution was faster than that in the HF-HNO_3-H_2O solution, with a consistent molarity of NO_3^-.

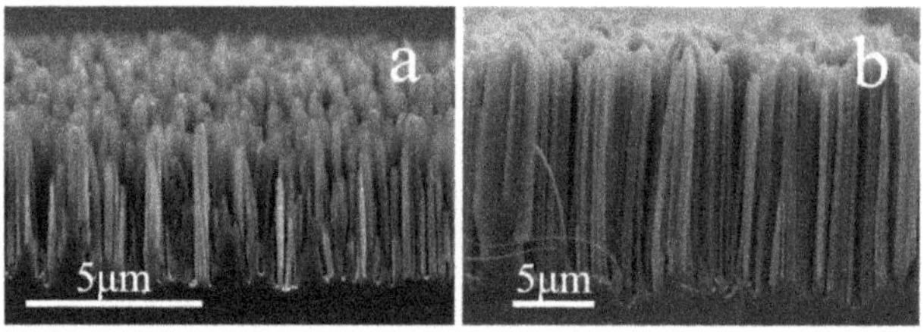

Figure 3. The SEM cross-section morphology of silver-plated SiNWs prepared in the etching system of 4.60 M HF and $M(NO_3)_x$: (**a**) 0.48 M HNO_3, (**b**) 0.16 M $Fe(NO_3)_3$.

3.2. Metal Ions' Dynamic Effect Analysis

Based on this experimental phenomenon, it was believed that NO_3^- also had an oxidizing effect in addition to Fe^{3+}. To discuss the influence of metal cations and expand the system range of the etching solution, the fourth-period element corresponding nitrate ($M(NO_3)_x$) and HNO_3 in the periodic table of elements were researched in this section in the hope that each $M(NO_3)_x$ would probably represent each main group element's corresponding $M(NO_3)_x$. Considering the metal elements in the fourth period of the periodic table, KNO_3, $Al(NO_3)_3$, $Cr(NO_3)_3$, $Mn(NO_3)_2$, $Ni(NO_3)_2$, $Co(NO_3)_2$, HNO_3, and $Ca(NO_3)_2$ were finally selected to form the etching solution with HF. After the reaction was completed, the cross-sectional morphology of the silicon wafer was observed with SEM, and the crystal orientation of the silicon nanostructure was observed with TEM.

The cross-sectional view of the SEM in Figure 4 shows that the silver nanoparticle layer is the catalyst, and the etching solution containing $M(NO_3)_x$ and HF can achieve regular SiNW arrays on the surface of a single-crystal silicon wafer. Most of the arrays are dense, one-dimensional arrays with a line diameter of 100–200 nm. However, the lengths of the SiNW arrays are different in different $M(NO_3)_x$ etching solutions, which means that the etching rate of silicon is also significantly different in different $M(NO_3)_x$ etching solutions. After one hour of etching reaction at 50 °C, the length of the SiNW arrays prepared in KNO_3 and $Al(NO_3)_3$ etching solution was about 3 μm (Figure 4a,b). The length of the SiNW arrays prepared in $Mn(NO_3)_2$ and $Cr(NO_3)_3$ etching solutions was the shortest, about 1 μm (Figure 4c,d). The lengths of the SiNW arrays prepared in $Ni(NO_3)_2$, $Co(NO_3)_3$, and $Ca(NO_3)_3$ etching solutions were about 5–6 μm (Figure 4e–g). As shown in Figure 4h, observing the microscopic surface morphology of a single silicon wire with a high-power TEM can find that the diameter of a single silicon wire is between 100 and 200 nm, which is also the (100) crystal orientation.

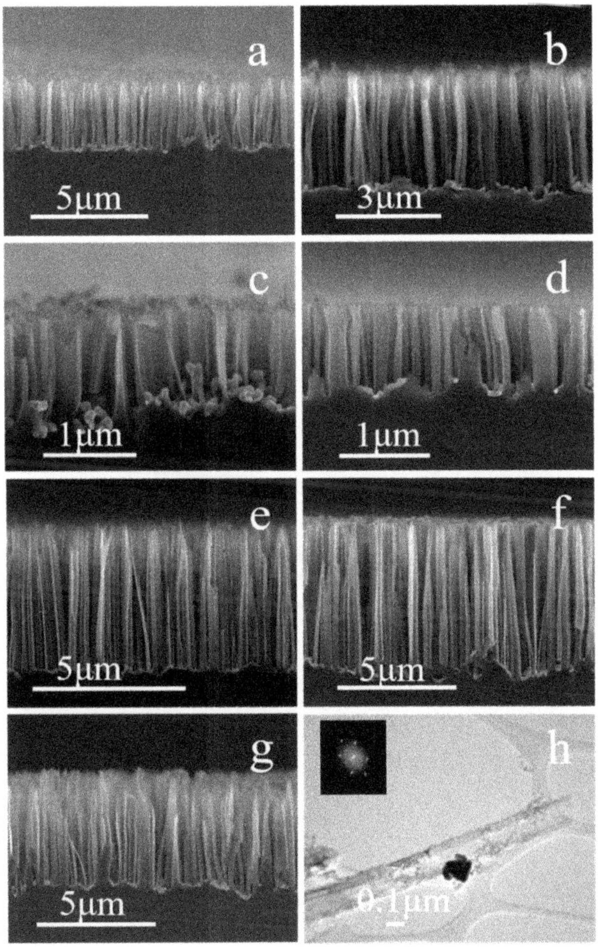

Figure 4. The surface micro and nano structure morphology of the silicon wafer etched in the $M(NO_3)_x$ and 4.6 M HF etching solution for one hour: (**a**) 0.05 M KNO_3, (**b**) 0.16 M $Al(NO_3)_3$, (**c**) 0.24 M $Mn(NO_3)_2$, (**d**) 0.16 M $Cr(NO_3)_3$, (**e**) 0.24 M $Ni(NO_3)_2$, (**f**) 0.24 M $Co(NO_3)_2$, (**g**) 0.24 M $Ca(NO_3)_2$, (**h**) TEM of one piece of silicon nanowire etched in etching solution of 0.05 M KNO_3.

The above results show that in different $M(NO_3)_x$ systems, the etching rate is also different, although the NO_3^- is an oxidant that continuously promotes the etching reaction. However, when only the types of metal ions in the $M(NO_3)_x$ are different, there are still some differences in the SiNW arrays. Therefore, metal ions are still one of the factors that dynamically affect the etching process. Considering the strong complexing ability of F^- in the etching system, the types of ions are very complex, the toxicity of HF is strong, and the characterization methods of ion species are limited. It was also found that cation concentration has a great influence on the etching results. KNO_3 was selected to research the effect of metal ions on the etching reaction, and the influence of different concentrations of KNO_3 on the system was also investigated.

Keeping the concentration of HF unchanged, when using 0.48 M KNO_3 etching solution, the surface of the silicon wafer was gray-white after reacting at 50 °C for three hours, and, observed under an SEM, as shown in Figure 5a, there was a layer of granular material attached. As shown in the cross-section shown in Figure 5b, the silicon wafer was not etched. The essence of MACE is that the galvanic cell drives the reaction of silicon in an HF solution system with an oxidant. The cathode and anode, the electrolyte, and the connected circuit are indispensable. During the etching reaction, K^+ will combine with F^- to form an insoluble K_2SiF_6 film that is closely attached to the surface of the silicon wafer, preventing the etching solution from contacting the silicon wafer and thereby inhibiting the occurrence of etching. With 0.05 M KNO_3, after the F^- was completely complexed with the K^+, there was still a large amount of residue, which was enough to support the occurrence of subsequent etching reactions. The SiNW arrays are shown in Figure 5c,d. When adding 0.48 M KNO_3, most of the F^- ions in the system were complexed by K^+ and formed insoluble K_2SiF_6, which adhered to the surface of the silicon wafer and inhibited the occurrence of the etching reaction. To verify the crystal composition of the gray-white substance on the surface of the silicon wafer, X-ray diffraction was used to characterize it. As shown in Figure 5f, the positions of the peaks in the XRD pattern correspond exactly to those of standard K_2SiF_6 (the power diffraction file 52-1831). Therefore, the small particles on the SiNW array side wall shown in Figure 5e are K_2SiF_6.

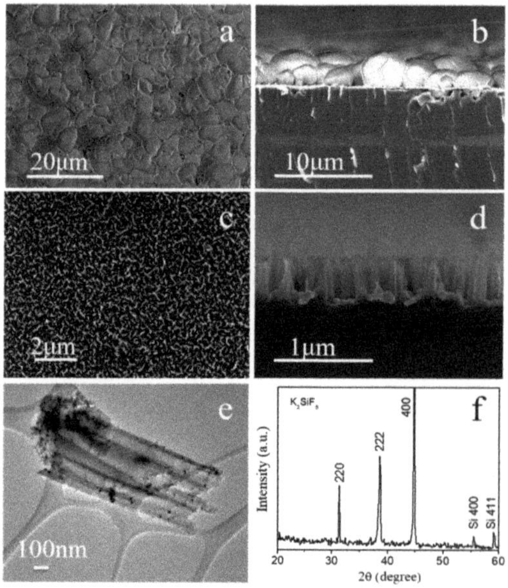

Figure 5. The morphology and structure analyses of silicon wafer coated with a silver nanoparticle layer and reacted in different concentrations of KNO_3 and 4.6 M HF etching solutions: (**a**) 0.40 M

KNO_3, etching for 3 h, (**b**) cross-sectional view of Figure 5a, (**c**) 0.05 M KNO_3, etching for 30 min, (**d**) cross-section of Figure 5c, (**e**) TEM of one silicon nanowire etched in 0.05 M KNO_3 for 30 min, (**f**) XRD of sample etched in 0.05 M KNO_3 for 30 min.

3.3. Mechanism Analysis

Based on the above experimental results, the influence of Fe^{3+} and K^+ in solution is further understood, and its mechanism is now updated. As shown in Figure 6, during the etching process, NO_3^- and Fe^{3+} are both oxidants; as oxidated silicon is dissolved by HF, the silver particles gradually sink as the reaction proceeds, while keeping dissolution and deposition in balance at the same time. In the HF-KNO_3-H_2O system, as the concentration of KNO_3 in the system increases, a layer of K_2SiF_6 will be formed on the surface of the silicon wafer, which hinders the continuation of the reactions. According to our previous work [10], it is suggested that the anodic dissolution of silicon during MACE of silicon follows both the divalent and tetravalent dissolution processes. The cathode is the reduction of the oxidant. Although Fe^{3+} and NO_3^- are cations and anions, respectively, both Fe^{3+} and NO_3^- are oxidizing in the etching solution; as they are all reacting at the cathode, they attack the same sections.

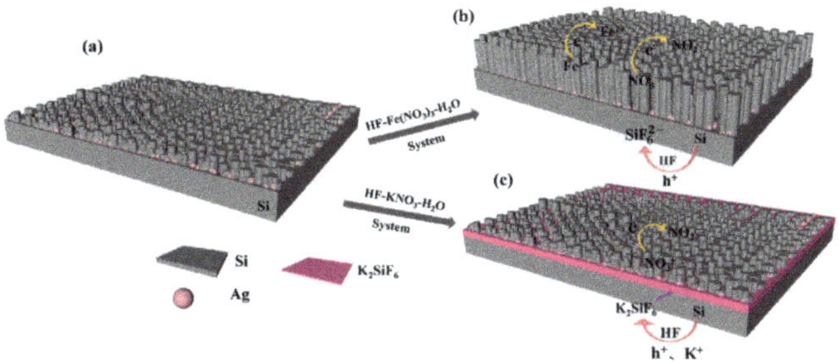

Figure 6. Schematic diagrams illustrating. (**a**) Silicon wafers coated with silver nanoparticle layer. (**b**) Silicon wafer coated with silver nanoparticle layer and reacted in HF-$Fe(NO_3)_3$-H_2O system. (**c**) Silicon wafer coated with silver nanoparticle layer and reacted in HF-KNO_3-H_2O system.

In the HF-$Fe(NO_3)_3$-H_2O system, Fe^{3+} is reduced to Fe^{2+}, and NO_3^- is reduced to NO_2^-. In the HF-KNO_3-H_2O system, most of the F^- ions in the system are complexed by K^+ and form insoluble K_2SiF_6, which adheres to the surface of the silicon wafer and inhibits the occurrence of etching reactions. The anodic and cathodic half-cell reactions in HF-$Fe(NO_3)_3$-H_2O and HF-KNO_3-H_2O system are described as follows [20,33,34]:

In the HF-$Fe(NO_3)_3$-H_2O system (as shown in Figure 6a)

Anodic:

(1) $Si + 2F^- + 2h^+ \rightarrow SiF_2$

$SiF_2 + 4F^- + 2H^+ \rightarrow SiF_6^{2-} + H_2$

(2) $Si + 2H_2O + 4h^+ \rightarrow SiO_2 + 4H^+$

$SiO_2 + 6HF \rightarrow 2H^+ + SiF_6^{2-} + 2H_2O$

Cathodic:

$NO_3^- + 2H^+ + 2e^- \rightarrow NO_2^- + H_2O$

$Fe^{3+} + e^- \rightarrow Fe^{2+}$

In the HF-KNO_3-H_2O system (as shown in Figure 6b)

Anodic:

(1) $Si + 2F^- + 2h^+ \rightarrow SiF_2$

$SiF_2 + 4F^- + 2K^+ + 2H^+ \rightarrow K_2SiF_6 + H_2$

(2) $Si + 2H_2O + 4h^+ \rightarrow SiO_2 + 4H^+$

$SiO_2 + 2K^+ + 6HF \rightarrow 2H^+ + K_2SiF_6 + 2H_2O$

Cathodic:

$NO_3^- + 2H^+ + 2e^- \rightarrow NO_2^- + H_2O$

4. Conclusions

In this paper, the oxidative roles of Fe^{3+} and NO_3^- during the MACE of silicon were proved, and metal ions' dynamic effect on MACE of silicon in the HF-M$(NO_3)_x$-H_2O etching acid solution (where M(NO_3)x is: KNO_3, $Al(NO_3)_3$, $Cr(NO_3)_3$, $Mn(NO_3)_2$, $Ni(NO_3)_2$, $Co(NO_3)_2$, HNO_3, and $Ca(NO_3)_2$) were discussed. The results showed that the type of cation affects the morphology of SiNW arrays and the rate of etching. Crystal composition analysis revealed that the influence of cation type and concentration on MACE is realized by the stability of the metal and its reaction products in the etching solutions. In HF-KNO_3-H_2O solution, the combination of K^+ and F^- affects the dissolution of SiO_2 or directly generates an insoluble substance attached to the wafer surface, thus inhibiting the etching of silicon. As the concentration of KNO_3 in the system increases, a layer of K_2SiF_6 is formed, which hinders the continuation of the etching reactions. This work is further helpful in revealing the metal ions' dynamic effect on MACE of silicon in the HF-M$(NO_3)_x$-H_2O etching acid solution, paves the way for further study of the mechanism of MACE, and provides a basis for us to find convenient and low-cost MACE methods.

Author Contributions: Conceptualization, X.Y.; methodology, X.Y. and Y.L.; formal analysis, X.Y. and B.Z.; investigation, X.Y. and Y.L.; resources, L.W. and Y.H.; writing—original draft preparation, B.Z. and Z.L.; writing—review and editing, X.Y., Y.L., T.T., Y.W. and Y.H.; supervision, L.W. and Y.H. All authors have read and agreed to the published version of the manuscript.

Funding: This work was supported by the Key Research and Development Program of Hubei Province (Grant No. 2023BAB122), the National Natural Science Foundation of China (Grant No. 61904130), and the Key Laboratory of Hubei Province for Coal Conversion and New Carbon Materials (Grant No. WKDM201907).

Institutional Review Board Statement: Not applicable.

Informed Consent Statement: Not applicable.

Data Availability Statement: Data is contained within the article.

Acknowledgments: We would like to thank Wang at the Analytical & Test Center of Wuhan University of Science and Technology for their help with SEM analysis.

Conflicts of Interest: The authors declare no conflict of interest.

References

1. Dhanabalan, A.; Song, B.F.; Biswal, S.L. Extreme Rate Capability Cycling of Porous Silicon Composite Anodes for Lithium-Ion Batteries. *ChemElectroChem* **2021**, *8*, 3318–3325. [CrossRef]
2. Nulu, A.; Nulu, V.; Sohn, K.Y. Si/SiOx Nanoparticles Embedded in a Conductive and Durable Carbon Nanoflake Matrix as an Efficient Anode for Lithium-Ion Batteries. *ChemElectroChem* **2020**, *7*, 4055–4065. [CrossRef]
3. Ulvestad, A.; Reksten, A.H.; Andersen, H.F.; Carvalho, P.A.; Jensen, I.J.T.; Nagell, M.U.; Mæhlen, J.P.; Kirkengen, M.; Koposov, A.Y. Crystallinity of Silicon Nanoparticles: Direct Influence on the Electrochemical Performance of Lithium Ion Battery Anodes. *ChemElectroChem* **2020**, *7*, 4349–4353. [CrossRef]
4. Wang, S.; Liu, H.; Han, J. Comprehensive Study of Au Nano-Mesh as a Catalyst in the Fabrication of Silicon Nanowires Arrays by Metal-Assisted Chemical Etching. *Coatings* **2019**, *9*, 149. [CrossRef]
5. Prigoda, K.; Ermina, A.; Bolshakov, V.; Nazarov, D.; Ezhov, I.; Lutakov, O.; Maximov, M.; Tolmachev, V.; Zharova, Y. The Array of Si Nanowires Covered with Ag Nanoparticles by ALD: Fabrication Process and Optical Properties. *Coatings* **2022**, *12*, 1748. [CrossRef]
6. Shanshan, W.; Yan, Z. Effect of High-Temperature Annealing on Raman Characteristics of Silicon Nanowire Arrays. *Coatings* **2023**, *13*, 793. [CrossRef]
7. Chen, Q.; Zhou, G.; Zhu, J.; Fan, C.; Li, X.G.; Zhang, Y. Ultraviolet Light Emission from Porous Silicon Hydrothermally Prepared. *Phys. Lett. A* **1996**, *224*, 133–136. [CrossRef]
8. Chen, Q.; Zhu, J.; Li, X.-G.; Fan, C.G.; Zhang, Y.H. Photoluminescence in Porous Silicon Obtained by Hydrothermal Etching. *Phys. Lett. A* **1996**, *220*, 293–296. [CrossRef]
9. Peng, K.Q.; Hu, J.J.; Yan, Y.J.; Wu, Y.; Fang, H.; Xu, Y.; Lee, S.T.; Zhu, J. Fabrication of Single-Crystalline Silicon Nanowires by Scratching a Silicon Surface with Catalytic Metal Particles. *Adv. Funct. Mater.* **2005**, *16*, 387–394. [CrossRef]
10. Hu, Y.; Peng, K.-Q.; Qiao, Z.; Huang, X.; Zhang, F.-Q.; Sun, R.-N.; Meng, X.-M.; Lee, S.-T. Metal-Catalyzed Electroless Etching of Silicon in Aerated HF/H_2O Vapor for Facile Fabrication of Silicon Nanostructures. *Nano Lett.* **2014**, *14*, 4212–4219. [CrossRef]

11. Jiang, W.; Ya, H.; Haichuan, Z.; Haoxin, F.; Yachun, W.; Chenliang, H.; Kui-Qing, P. Oxidant Concentration Modulated Metal/Silicon Interface Electrical Field Mediates Metal-Assisted Chemical Etching of Silicon. *Adv. Mater. Interfaces* **2018**, *5*, 1801132.
12. Peng, K.; Lu, A.; Zhang, R.; Lee, S.-T. Motility of Metal Nanoparticles in Silicon and Induced Anisotropic Silicon Etching. *Adv. Funct. Mater.* **2008**, *18*, 3026–3035. [CrossRef]
13. Hu, Y.; Jin, C.; Liu, Y.; Yang, X.; Liao, Z.; Zhang, B.; Zhou, Y.; Chen, A.; Wu, L.; Liu, J.; et al. Metal Particle Evolution Behavior during Metal Assisted Chemical Etching of Silicon. *ECS J. Solid State Sci. Technol.* **2021**, *10*, 084002. [CrossRef]
14. Peng, K.Q.; Yan, Y.J.; Gao, S.P.; Zhu, J. Synthesis of Large-Area Silicon Nanowire Arrays via Self-Assembling Nanoelectrochemistry. *Adv. Mater.* **2002**, *14*, 1164–1167. [CrossRef]
15. To, W.-K.; Tsang, C.-H.; Li, H.-H.; Huang, Z. Fabrication of N-Type Mesoporous Silicon Nanowires by One-Step Etching. *Nano Lett.* **2011**, *11*, 5252–5258. [CrossRef]
16. Leng, X.; Wang, C.; Yuan, Z. Progress in Metal-Assisted Chemical Etching of Silicon Nanostructures. *Procedia CIRP* **2020**, *89*, 26–32. [CrossRef]
17. Venkatesan, R.; Arivalagan, M.K.; Venkatachalapathy, V.; Pearce, J.M.; Mayandi, J. Effects of Silver Catalyst Concentration in Metal Assisted Chemical Etching of Silicon. *Mater. Lett.* **2018**, *221*, 206–210. [CrossRef]
18. Aca-López, V.; Quiroga-González, E.; Gómez-Barojas, E.; Świątowska, J.; Luna-López, J.A. Effects of the Doping Level in the Production of Silicon Nanowalls by Metal Assisted Chemical Etching. *Mater. Sci. Semicond. Process.* **2020**, *118*, 105206. [CrossRef]
19. Han, H.; Huang, Z.; Lee, W. Metal-Assisted Chemical Etching of Silicon and Nanotechnology Applications. *Nano Today* **2014**, *9*, 271–304. [CrossRef]
20. Hu, Y.; Fu, H.; Wang, J.; Sun, R.; Wu, L.; Liu, Y.; Xu, J.; Liu, J.; Peng, K.-Q. Carbon Induced Galvanic Etching of Silicon in Aerated HF/H_2O Vapor. *Corros. Sci.* **2019**, *157*, 268–273. [CrossRef]
21. Hassan, H.H.; Chazalviel, J.N.; Neumann-Spallart, M.; Ozanam, F.; Etman, M. Chemical Limitations to the Anodic Dissolution of P-Si in Fluoride Media in the Presence of Alkali Metal Cations. *J. Electroanal. Chem.* **1995**, *381*, 211–214. [CrossRef]
22. Hassan, H.H.; Fotouhi, B.; Sculfort, J.L.; Abdel-Rehiem, S.S.; Etman, M.; Ozanam, F.; Chazalviel, J.N. Effect of Alkali-Metal and Some Quaternary-Ammonium Cations on the Anodic Dissolution of p-Si in Fluoride Media. *J. Electroanal. Chem.* **1996**, *407*, 105–113. [CrossRef]
23. Nahm, K.S.; Seo, Y.H.; Lee, H.J. Formation Mechanism of Stains during Si Etching Reaction in HF–Oxidizing Agent–H_2O Solutions. *J. Appl. Phys.* **1997**, *81*, 2418–2424. [CrossRef]
24. Hun, S.Y.; Suk, N.K.; Bang, L.K. Mechanistic Study of Silicon Etching in HF-$KBrO_3$-H_2O Solution. *J. Electrochem. Soc.* **1993**, *140*, 1453.
25. Lin, H.; Wu, F.; Gao, P.; Shen, W. Shape-Controlled Silicon Microwire Arrays from Au–Ag-Catalyzed Metal-Assisted Chemical Etching for Radial Junction Solar Cells. *ACS Appl. Energy Mater.* **2019**, *2*, 5871–5876. [CrossRef]
26. Booker, K.; Rahman, S.; Chong, T.-K.; Mankelow, R.; Weber, K.; Blakers, A. A Robust Metal-Assisted Etching Process for Ag-Catalyzed Texturing of Silicon. *IEEE J. Photovolt.* **2015**, *5*, 766–773. [CrossRef]
27. Shaoyuan, L.; Wenhui, M.; Yang, Z.; Xiuhua, C.; Yongyin, X.; Mingyu, M.; Wenjie, Z.; Feng, W. Fabrication of Porous Silicon Nanowires by MACE Method in HF/H_2O_2/$AgNO_3$ System at Room Temperature. *Nanoscale Res. Lett.* **2014**, *9*, 196. [CrossRef]
28. Milazzo, R.G.; D'arrigo, G.; Spinella, C.; Grimaldi, M.G.; Rimini, E. Ag-Assisted Chemical Etching of (100) and (111) n-Type Silicon Substrates by Varying the Amount of Deposited Metal. *J. Electrochem. Soc.* **2012**, *159*, D521. [CrossRef]
29. Tian, B.; Zhang, J. Morphology-Controlled Synthesis and Applications of Silver Halide Photocatalytic Materials. *Catal. Surv. Asia* **2012**, *16*, 210–230. [CrossRef]
30. Bi, Y.; Ye, J. In Situ Oxidation Synthesis of Ag/AgCl Core–Shell Nanowires and Their Photocatalytic Properties. *Chem. Commun.* **2009**, *2009*, 6551–6553. [CrossRef]
31. Bi, Y.; Ye, J. Direct Conversion of Commercial Silver Foils into High Aspect Ratio AgBr Nanowires with Enhanced Photocatalytic Properties. *Chem.—A Eur. J.* **2010**, *16*, 10327–10331. [CrossRef] [PubMed]
32. Bi, Y.; Ye, J. Heteroepitaxial Growth of Platinum Nanocrystals on AgCl Nanotubes via Galvanic Replacement Reaction. *Chem. Commun.* **2010**, *46*, 1532. [CrossRef] [PubMed]
33. De Groot, M.; Koper, M.T. The Influence of Nitrate Concentration and Acidity on the Electrocatalytic Reduction of Nitrate on Platinum. *J. Electroanal. Chem.* **2004**, *562*, 81–94. [CrossRef]
34. Dima, G.E.; Beltramo, G.L.; Koper, M.T.M. Nitrate Reduction on Single-Crystal Platinum Electrodes. *Electrochim. Acta* **2005**, *50*, 4318–4326. [CrossRef]

coatings

MDPI

Article

Controlled Aggregation of Cobalt and Platinum Atoms via Plasma Treatment for Exceptional Hydrogen Evolution Reaction Activity

Guoqing Zhang [1], Jiankun Li [1], Yixing Wang [1,2], Linfeng Lei [1,2] and Linzhou Zhuang [1,*]

[1] State Key Laboratory of Chemical Engineering, East China University of Science and Technology, Shanghai 200237, China; zgq18482573871@163.com (G.Z.); jiankunli@mail.ecust.edu.cn (J.L.); wangyx@szlab.ac.cn (Y.W.); linfeng.lei@ecust.edu.cn (L.L.)

[2] Suzhou Laboratory, Suzhou 215000, China

* Correspondence: lzzhuang@ecust.edu.cn

Abstract: Designing and developing highly active, stable, and cost-effective hydrogen evolution reaction (HER) catalysts is crucial in the field of water electrolysis. In this study, we utilize N-doped porous carbon (CoNC) derived from zeolite imidazole metal–organic frameworks (ZIF-67) as support and prepare CoNC-Pt-IM-P via chemical impregnation (CoNC-Pt-IM) and plasma treatment. Systematic analyses reveal that calcined CoNC with pyridinic nitrogen could serve as a robust support to strongly anchor PtCo nanoclusters, while argon plasma treatment could lead to a noticeable aggregation of Co and Pt atoms so as to alter the electronic environment and enhance intrinsic HER catalytic activity. CoNC-Pt-IM-P could exhibit outstanding catalytic activity toward HER, achieving an exceptionally low overpotential of 31 mV at the current density of -10 mA cm^{-2} and a Tafel slope of 36 mV dec^{-1}. At an overpotential of 50 mV, its mass activity reaches 4.90 A mg_{Pt}^{-1}, representing enhancements of 1.5 times compared to CoNC-Pt-IM and 12.3 times compared to commercial 20 wt% Pt/C. Furthermore, it could operate stably for over 110 h at a current density of -10 mA cm^{-2}, demonstrating its exceptional durability. This work uses plasma treatment to achieve the controllable aggregation of Co and Pt atoms to enhance their catalytic activity, which has the advantage of avoiding excessive particle aggregation compared to the commonly used method of high-temperature calcination.

Keywords: hydrogen evolution reaction; plasma; intermetallic compounds; controlled aggregation

1. Introduction

Hydrogen, as a clean and sustainable energy carrier, possesses the highest energy density among known fuels [1–4]. When utilized in fuel cells, it not only offers extremely high energy conversion efficiency but also produces water as the sole product, rendering it a zero-pollution energy source. Therefore, the development of water electrolysis technologies to convert renewable energy into hydrogen is crucial for achieving sustainable energy in the future [5,6]. Significant progress has been made in hydrogen production via water electrolysis and its conversion to electricity through fuel cells. Proton exchange membrane water electrolysis (PEMWE) technology, in particular, stands out due to its superior current density compared to other water electrolysis techniques [7–9]. To reduce the energy required for hydrogen production through electrolysis, efficient and stable catalysts are essential to lower the activation energy of the reaction. To date, noble metal catalysts, mainly Pt-based catalysts, are the most effective for the hydrogen evolution reaction (HER) in acidic environments [10,11]. However, state-of-the-art Pt/C catalysts, which typically contain 20 wt% Pt, face limitations in application due to issues including the high cost of Pt, the poor utilization efficiency of Pt, and unsatisfactory stability. Specifically, poor stability mainly stems from Pt particle aggregation, carbon support corrosion, and Pt dissolution in the electrolyte. Therefore, designing and synthesizing Pt-based catalysts with a low noble metal content, high activity, and enhanced stability is crucial. Current optimization strategies to address these challenges include single-atom catalysts, alloying, and combining Pt with carbon-based supports [12,13].

In recent years, metal–organic frameworks (MOFs) have garnered significant attention due to their simple synthesis methods, large surface area, high porosity, and strong loading capacity [14–16]. MOFs have been widely used to prepare heteroatom-doped porous carbon as sacrificial template precursors to enhance the catalytic performance of alloy catalysts supported on carbon [17,18]. For instance, Li et al. encapsulated Co@Ir core–shell nanoparticles (NPs) in N-doped porous carbon derived from zeolite imidazole metal–organic frameworks (ZIF-67) and tested its HER activity in 1.0 M KOH. They observed an overpotential of 121 mV at a current density of 10 mA cm^{-2}. In their Co@Ir/NC-10% catalyst, the Ir shell provided abundant active sites for the oxygen evolution reaction (OER), while the Co-NC framework offered highly active sites for HER [19]. Similarly, Yang et al. developed Co@Pd nanoclusters (NC) through the controlled pyrolysis of ZIF-67, surpassing Pd/C as an HER catalyst and delivering a current density of 10 mA cm^{-2} at an overpotential of 98 mV with a Tafel slope of 55 mV dec^{-1}. Furthermore, they fabricated a CoPt-platinum atomic site (PtSA)/nitrogen-doped porous carbon framework (NDPCF) electrocatalyst, which exhibited ultralow overpotentials under both alkaline and acidic conditions at a high current density of -200 mA cm^{-2} (over potential: 110 mV in acidic conditions), demonstrating promising long-term durability for up to 100 h or 10,000 cycles. This performance was attributed to the synergistic effects of the PtSA and CoPt alloy [20]. However, comprehensive insights into tuning the microenvironment of catalysts, such as adjusting neighboring Pt atomic sites to enhance the electrocatalytic activity and stability of PtCo nanoclusters, are still lacking.

This study demonstrates that plasma bombardment can effectively modulate the electronic structure of Pt and Co atoms encapsulated within NC nanoframeworks derived from high-temperature (800 °C) pyrolysis. This results in a stable framework with nitrogen vacancies that could securely anchor Pt and Co atoms. After plasma treatment, the Pt and Co atoms are more tightly bonded, while the abundant nitrogen sites provide ample electrons, significantly enhancing the catalyst's activity and corrosion resistance under acidic conditions. In a 0.5 M H_2SO_4 solution, the CoNC-Pt-IM-P catalyst with a Pt loading of just 2.3 wt% exhibited excellent catalytic performance with overpotentials of only 31 mV at -10 mA cm^{-2} and 98 mV at -200 mA cm^{-2}. Consequently, its mass activity at an overpotential of 50 mV reached a value as high as 4.9 A $mgPt^{-1}$. Additionally, CoNC-Pt-IM-P can stably operate for over 110 h at -10 mA cm^{-2}, demonstrating exceptional catalytic stability. This work presents a simple and practical strategy for synthesizing multi-metallic catalysts, which can be extended to the preparation of other bi-metallic or multi-metallic catalysts, offering valuable insights for future research.

2. Experimental Procedure

2.1. Chemicals

2-methylimidazole (2-MeI, Aladdin, AR), cobalt(II) nitrate hexahydrate ($Co(NO_3)_2 \cdot 6H_2O$, Aladdin, Shanghai, China, AR), cobalt(II) phthalocyanine (CoPC, J&K Scientific, Beijing, China, 92%), dicyandiamide (DCDA, Energy Chemical, Shanghai, China, 99%), chloroplatinic acid hexahydrate ($H_2PtCl_6 \cdot 6H_2O$, Aladdin, AR, Pt $\geq$ 37.5%), isopropyl alcohol (C_3H_8O, Macklin, Shanghai, China, 99.5%), Nafion (LIGE SCIENCE, Tianjin, China, 5%), carbon paper (TGP-H-060, Sinero, Suzhou, China), Pt/C (HWRK CHEM, Beijing, China, 20 wt%), deionized water, and sulfuric acid (H_2SO_4, Sinopharm Chemical Reagent Co., Ltd., Shanghai, China, 98%) were utilized in this study.

2.2. Synthesis of Electrocatalysts

First, 2.46 g of 2-methylimidazole (2-MeI) and 2.18 g of $Co(NO_3)_2 \cdot 6H_2O$ were dissolved in 30 mL of methanol. After complete dissolution, the methanol solution of $Co(NO_3)_2$ was slowly added to the 2-Mel solution while stirring. Once the addition was completed, the mixture was sealed in a beaker and stirred for 12 h. The resulting mixture was then filtered, washed three times, and dried in a vacuum oven at room temperature for 6 h, yielding a uniform purple ZIF-67 powder.

To prepare CoNC, ZIF-67 was placed in a tubular furnace that was filled with argon as a protective atmosphere. It was heated to 800 °C at a rate of 5 °C min^{-1} and maintained at this temperature for two hours. After cooling, the product was ground to obtain a fine black powder, resulting in the N-doped porous carbon material, which was denoted as CoNC.

The preparation of CoNC-Pt-IM involved adding 100 mg of CoNC support powder to 20 mL of deionized water, followed by the addition of 5 mL of $H_2PtCl_6{\cdot}6H_2O$ solution (0.83 mg Pt mL^{-1}). The mixture was stirred for 5 h and then filtered, washed three times, and dried in a vacuum oven at room temperature for 6 h. The resulting product was ground to obtain a uniform CoNC-Pt-IM.

The preparation of the control sample, ZIF-67-Pt, followed the same protocol used for CoNC-Pt-IM, but the ZIF-67 was used as the support instead of CoNC, while all other steps remained unchanged.

For the preparation of CoNC-Pt-IM-P, ZIF-67 was first calcined to obtain the support, followed by impregnation with the Pt precursor solution. Subsequently, plasma treatment was applied using argon plasma at a power setting of 100 W for a duration of 10 min. This process yielded the final product, CoNC-Pt-IM-P.

2.3. Characterizations

X-ray Powder Diffraction (XRD) data were collected on a Bruker D8 Advance X-ray Poly.crystalline Diffractometer at 40 kV/40 mA with Cu Ka radiation (k = 1.541874 Å) in the angular range of 5–80° for the 2θ angle. X-ray photoemission spectroscopy (XPS, Al-K-alpha, Thermo Scientific, Waltham, MA, USA) and scanning electron microscopy (SEM, Jeol, Tokyo, Japan) were carried out on carbon tape. Transmission electron microscopy (TEM, Jeol) measurements were completed on a carbon-coated copper TEM grid support.

2.4. Electrochemical Measurements

All tests were conducted using an Autolab (Metrohm PGSTAT302N, Herisau, Switzerland) electrochemical workstation in a three-electrode configuration. A carbon rod and Hg/Hg_2SO_4/saturated K_2SO_4 electrodes were used as the counter and reference electrodes, respectively, and the working electrode was prepared by mixing the samples (including CoNC, CoNC-Pt-IM, CoNC-Pt-IM-P, and commercial 20 wt%Pt/C) with Nafion and drop-casting them onto carbon paper for testing at 25 °C. The catalyst preparation procedure was as follows: 10 mg of catalyst powder, 100 μL of Nafion solution (5 wt%), and 900 μL of isopropanol were mixed and sonicated for 30 min until a homogeneous suspension was obtained; then, 100 μL of the prepared slurry was drop-cast onto a piece of carbon paper (1×1.5 cm^2, with a coated area of 1×1 cm^2) and left to dry for 30 min until the solvent was evaporated. This resulted in a uniform layer of catalyst powder that adhered to the carbon paper with a catalyst loading of 1.0 mg cm^{-2}. All electrochemical results were reported with respect to the reversible hydrogen electrode (RHE). Linear sweep voltammetry (LSV) was performed by sweeping the potential at a rate of 5 mV s^{-1} over a range of −0.7 V to 0.0 V in 0.5 M H_2SO_4. Data were collected after stable cyclic voltammetry (CV) curves were obtained. The measured potentials against Hg/Hg_2SO_4 were converted to RHE using the following equation:

$$E_{(RHE)} = E_{Hg/Hg2SO4} + 0.059 \times pH + 0.65$$

where $E_{Hg/Hg2SO4}$ is the working potential. To evaluate the charge transfer properties of the electrode material, electrochemical impedance spectroscopy (EIS) measurements were performed. The parameters were set as follows: a frequency range from 0.01 Hz to 100 kHz with a voltage of −0.7 V and an amplitude of 5 mV.

3. Results and Discussion

3.1. Catalyst Morphology

The synthesis process of the samples is illustrated in Figure 1a. The porous N-doped carbon (CoNC) was obtained by calcining ZIF-67 and then impregnating it in the Pt precursor solution to yield CoNC-Pt-IM. The plasma treatment was applied to achieve the stable anchoring of CoNC-Pt-IM-P. The morphology of the synthesized materials was examined using scanning electron microscopy (SEM). As shown in Figure S1a,b, the synthesized ZIF-67 exhibited a smooth surface and a dodecahedral structure with an average size of approximately 400 nm, consistent with literature reports [21,22]. After calcination, the obtained CoNC exhibited a rough and wrinkled surface while retaining its dodecahedral shape, indicating that the carbon framework remained intact (Figures 1b,c and S1c,d). Moreover, the CoNC-Pt-IM maintained its dodecahedral morphology after impregnation, showing no significant changes to the rough surface compared to CoNC (Figure 1d,e) and no observable particle loading. In contrast, after directly immersing ZIF-67 in H_2PtCl_6 solution, ZIF-67-Pt experienced noticeable hydrolysis and structural collapse, and thus, its morphology became irregular (Figure S1e,f). This demonstrated the superior structural stability of the CoNC support obtained through calcination [23]. Additionally, CoNC exhibited distinct porous structures, which was expected to facilitate electrolyte penetration and allow ions to interact more effectively at the solid–liquid interface. Similarly, CoNC-Pt-IM-P retained the same configuration after the plasma treatment [24].

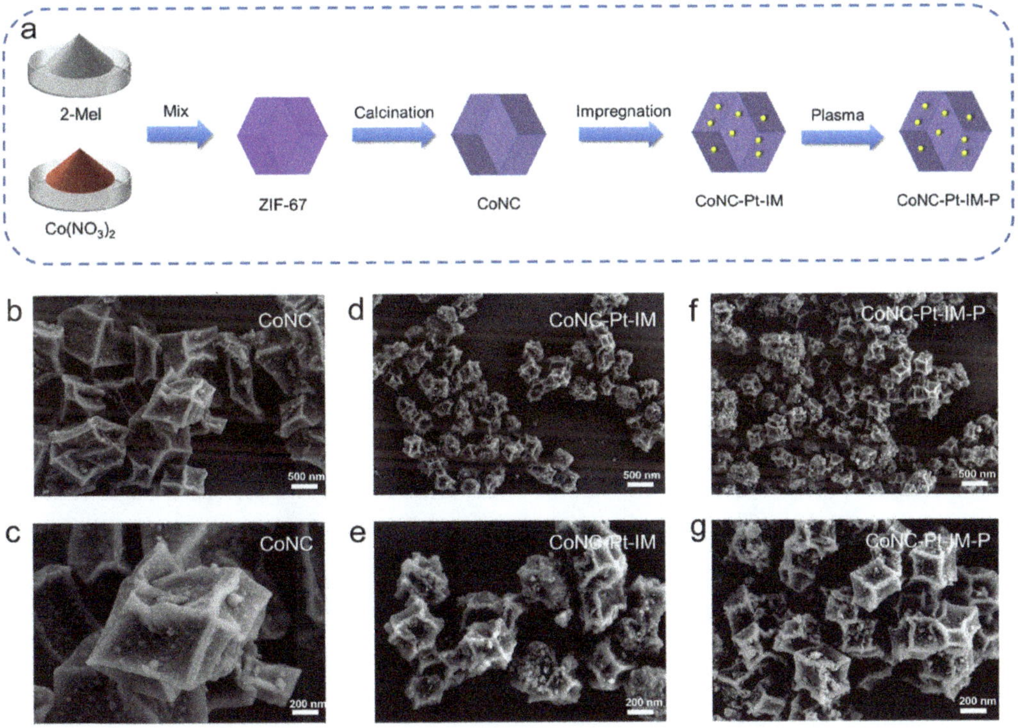

Figure 1. (**a**) A schematic diagram of the synthesis process of CoNC-Pt-IM-P. (**b**,**c**) An SEM image of CoNC. (**d**,**e**) An SEM image of CoNC-Pt-IM. (**f**,**g**) SEM images of CoNC-Pt-IM-P.

As shown in the transmission electron microscope (TEM) images (Figure 2a,b), the CoNC-Pt-IM support features uniformly distributed metal nanoparticles with an average particle size of 8.86 ± 5.26 nm. Subsequent measurements of the lattice spacings of the supported metal particles indicate values of 0.21 nm and 0.23 nm (Figure 2c–e), corresponding to the Co (111) and Pt (111) crystal planes, respectively [16,25]. This suggests the independent presence of both Co nanoparticles and Pt nanoparticles on the support, with a notable distance between them. High-Angle Annular Dark Field Scanning Transmission Electron Microscopy (HAADF-STEM) images display the elemental distribution (Figure S2a–d), further confirming a relatively uniform distribution of Pt atoms, which had no significant bonding to Co atoms. In contrast, CoNC-Pt-IM-P exhibits a distinct distribution of metal nanoparticles at a larger scale (Figure 2f), and the average particle size increases to 9.26 ± 4.82 nm (Figure 2g). Figure 2h,i reveal two neighboring metal nanoparticles with lattice spacings of 0.21 nm and 0.23 nm, corresponding to the Co (111) and Pt (111) crystal planes. Unlike the spatial separation observed in CoNC-Pt-IM, the CoNC-Pt-IM-P samples clearly show a closer proximity of the metal nanoparticles, indicating partial alloying. Additionally, as displayed in Figure 2j–m, compared to the uniform distribution of Pt in CoNC-Pt-IM, the Pt atoms in CoNC-Pt-IM-P show tighter integration with Co atoms. However, CoNC-Pt-IM and CoNC-Pt-IM-P both exhibit significant nitrogen doping with a relatively uniform distribution and no noticeable differences (Figure S3).

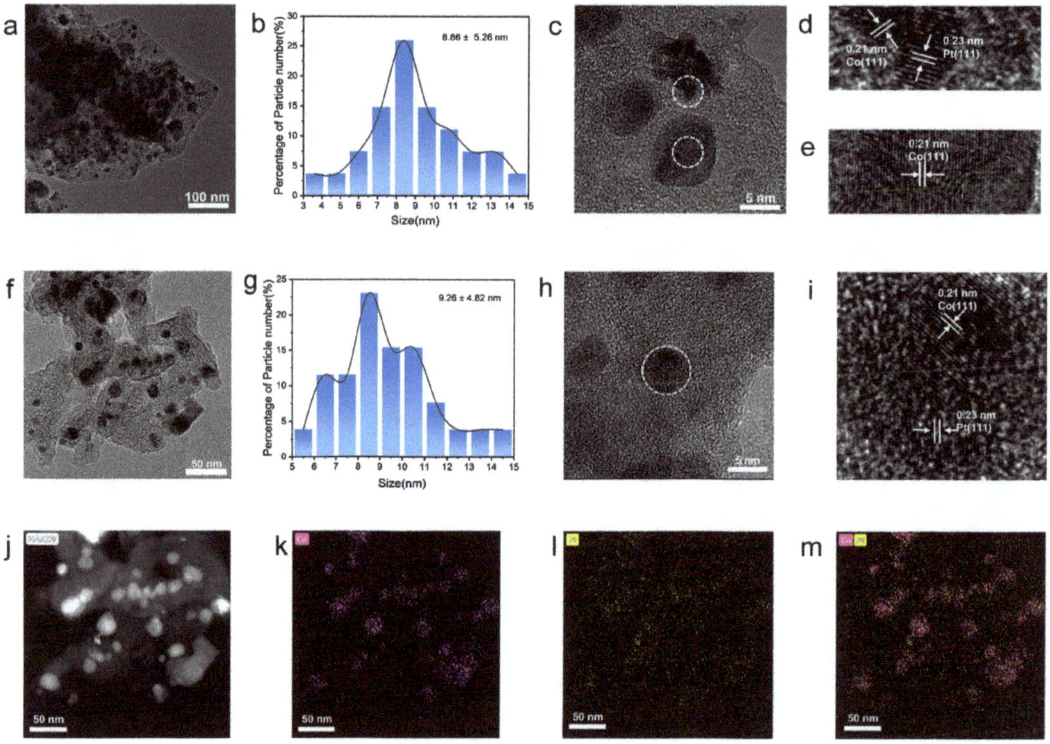

Figure 2. (**a**) TEM image of CoNC-Pt-IM. (**b**) Statistical analysis of particle size distribution for CoNC-Pt-IM. (**c**) Enlarged TEM image of CoNC-Pt-IM. (**d**,**e**) Lattice spacing measurements obtained from two selected areas in enlarged TEM image. (**f**) TEM image of CoNC-Pt-IM-P. (**g**) Statistical analysis of particle size distribution for CoNC-Pt-IM-P. (**h**) Enlarged TEM image of CoNC-Pt-IM-P. (**i**) Lattice spacing measurements from selected areas in enlarged image of CoNC-Pt-IM-P. (**j**–**m**) HAADF-STEM–EDS mapping of as-synthesized CoNC-Pt-IM-P.

3.2. Chemical Structure and Surface Composition

X-ray diffraction (XRD) data of the prepared samples are presented in Figure 3a. The XRD characteristic peaks of the synthesized MOF material matched well with the standard pattern of ZIF-67, confirming its successful synthesis (Figure S4a). However, after impregnation, the hydrolysis occurring during the impregnation process could have led to structure degradation so that the intensity of the characteristic peaks in the ZIF-67-Pt sample decreased significantly. This is consistent with the abovementioned SEM images (Figure S1e,f). As depicted in Figure 3a, after calcination, the characteristic peaks of ZIF-67 disappeared in the CoNC sample, with distinct diffraction peaks newly appearing at 44°, 51°, and 76°, which could be attributed to the metallic Co (PDF#15-0806), confirming the formation of Co nanoparticles [26]. The diffraction peaks of the CoNC-Pt-IM sample were nearly identical to those of CoNC, indicating that impregnation did not significantly alter the properties of the pristine support. This demonstrates the excellent stability of the carbon framework. After the plasma treatment, the intensity of the metallic Co diffraction peaks showed a significant increase in the XRD pattern of CoNC-Pt-IM-P, suggesting the growth of Co nanoparticles. Additionally, the peak at 52° should be attributed to the Co (200) plane. The absence of distinct crystalline peaks for Pt is likely due to the ultrasmall size and low content of Pt particles. Meanwhile, some of the original Co diffraction peaks had a slight shift, implying the partial alloying of PtCo.

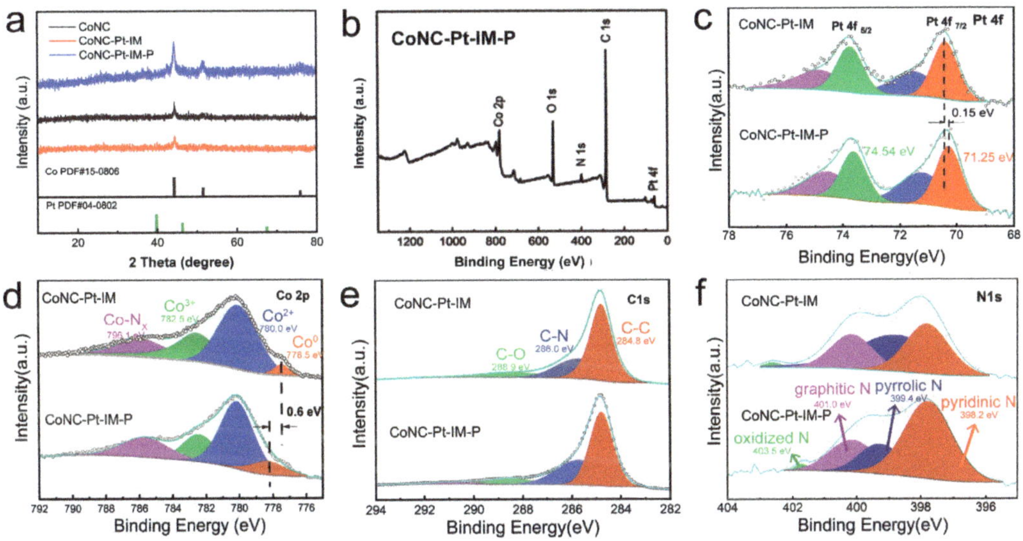

Figure 3. (**a**) The XRD pattern of the synthesized catalysts. (**b**) The XPS survey spectra and high-resolution XPS spectra: (**c**) Pt 4f, (**d**) Co 2p, (**e**) C 1s, and (**f**) N 1s of CoNC-Pt-IM-P, respectively.

To further elucidate the structural differences among the samples, X-ray photoelectron spectroscopy (XPS) was employed to investigate the surface chemical composition of these catalysts. The survey-scan XPS spectra of the catalysts (Figures 3b and S5) reveal peaks corresponding to Co 2p, Pt 4f, C 1s, and N 1s, with an additional peak around 520 eV attributed to O 1s from the adsorbed air on the catalyst surface. The Pt 4f spectrum of CoNC-Pt-IM-P shows two main peaks, Pt $4f_{7/2}$ (71.23 eV) and Pt $4f_{5/2}$ (72.45 eV), attributed to the Pt^0 and Pt^{2+} oxidation states (Figure 3c) [27,28]. Compared with CoNC-Pt-IM, the proportion of Pt^0 in CoNC-Pt-IM-P decreased from 55.3% to 52.3%, while Pt^{2+} increased from 44.6% to 47.7%, suggesting electronic transfer due to the partial alloying of PtCo. For comparison, in the Pt 4f spectrum of ZIF-67-Pt (Figure S6, Table S1), the peaks were mainly identified as Pt^{2+} (72.39 eV) and Pt^{4+} (75.74 eV), indicating that despite using the same impregnation method, the unique nature of the CoNC support was crucial for facilitating the reduction and anchoring of Pt during the process. The lower ratio of metallic Pt may have resulted in the decreased HER activity.

Furthermore, the high-resolution Co 2p spectra of CoNC-Pt-IM-P (Figure 3d, Table S2) displayed peaks corresponding to Co^0 (778.5 eV), Co^{2+} (780.0 eV), Co^{3+} (782.5 eV), and Co-N_x (796.1 eV) [29,30]. This indicated that cobalt species in CoNC-Pt-IM-P predominantly existed as cobalt nanoclusters featuring Co^{2+}-N coordination states. The C 1s spectra of CoNC-Pt-IM and CoNC-Pt-IM-P (Figures 3e and S6) both exhibited three peaks that were associated with graphitic C-C (284.8 eV), C-N (286.0 eV), and C-O [31,32]. In the N 1s spectra (Figure 3f), four peaks appeared at 398.2, 399.4, 401.0, and 403.5 eV, corresponding to pyridinic N, pyrrolic N, graphitic N, and oxidized N states, respectively [33]. Notably, the content of pyridinic nitrogen in CoNC-Pt-IM-P was higher than that in CoNC-Pt-IM (Figure S7d), suggesting that the PtCo alloy may be embedded within the CoNC framework.

3.3. *Electrochemical Activity Evaluation in Acidic Media*

To evaluate the hydrogen evolution reaction (HER) behavior, electrochemical tests were conducted on CoNC, CoNC-Pt-IM, CoNC-Pt-IM-P, and commercial Pt/C (20 wt%) in a 0.5 M H_2SO_4 aqueous solution. All potentials were referenced to the reversible hydrogen electrode (RHE) and corrected for 90% iR to eliminate the ohmic drop in the electrolyte. As shown in Figures 4a,b and S8, the HER activity of the pristine CoNC sample was poor, exhibiting a Tafel slope of 189 mV dec^{-1} and an overpotential of 161 mV at -10 mA cm^{-2}. After impregnation with Pt, CoNC-Pt-IM demonstrated significant improvement, with the Tafel slope and overpotential decreasing to 66 mV dec^{-1} and 33 mV, respectively, indicating the effective adhesion of Pt to the CoNC nanoframe and enhanced catalytic activity. For comparison, ZIF-67-Pt exhibited an overpotential of 109 mV at -10 mA cm^{-2} (Figure S7), both significantly inferior to CoNC-Pt-IM. The results highlight that the stable CoNC nanoframe could facilitate electrolyte penetration and enhance ion transfer at the solid–liquid interface. After subsequent plasma treatment, the HER catalytic performance of CoNC-Pt-IM-P further improved, achieving an ultralow overpotential of 31 mV at 10 mA cm^{-2} and a Tafel slope of 36 mV dec^{-1}, surpassing the commercial 20wt% Pt/C catalyst, which displayed an overpotential of 50 mV and a Tafel slope of 39 mV. Given that the CoNC-Pt-IM-P catalyst possessed a Pt loading of just 2.3 wt% (obtained from ICP-AES), the mass activity of CoNC-Pt-IM-P was calculated to be 4.9 A $mg_{Pt}{}^{-1}$, significantly higher than the 0.4 A $mg_{Pt}{}^{-1}$ for Pt/C.

The plasma treatment did not significantly alter the CoNC support or Pt loading amount, suggesting that it was the alloy formation that led to the significantly enhanced intrinsic catalytic activity. The electrochemical active surface area (ECSA) was characterized by measuring the double-layer capacitance (C_{dl}) of the catalysts within a non-Faradaic potential range. As illustrated in Figure S9, the C_{dl} of CoNC-Pt-IM-P (63.42 mF cm^{-2}) was greater than that of CoNC-Pt-IM (35.15 mF cm^{-2}) and CoNC (24.45 mF cm^{-2}), indicating that CoNC-Pt-IM-P provided more HER active sites. The electrochemical impedance spectroscopy (EIS) results (Figure 3d,e) reveal that the charge transfer resistance (R_{ct}) of CoNC-Pt-IM-P (0.47 Ω) was significantly lower than that of CoNC-Pt-IM (237.9 Ω) and CoNC (73.7 Ω), demonstrating its superior charge transfer capability. At -0.07 V, compared to the pristine CoNC (73.7 Ω), CoNC-Pt-IM-P displayed a much lower R_{ct} (0.47 Ω), indicating minimal interfacial resistance and a faster charge transfer process. Moreover, to meet industrial application requirements, HER catalysts should not only possess outstanding catalytic performance but also exhibit good stability. As shown in Figure 4e, the chronoamperometric tests indicated that CoNC-Pt-IM-P could stably operate for over 110 h at a current density of -10 mA cm^{-2}, demonstrating its excellent electrochemical stability [34,35].

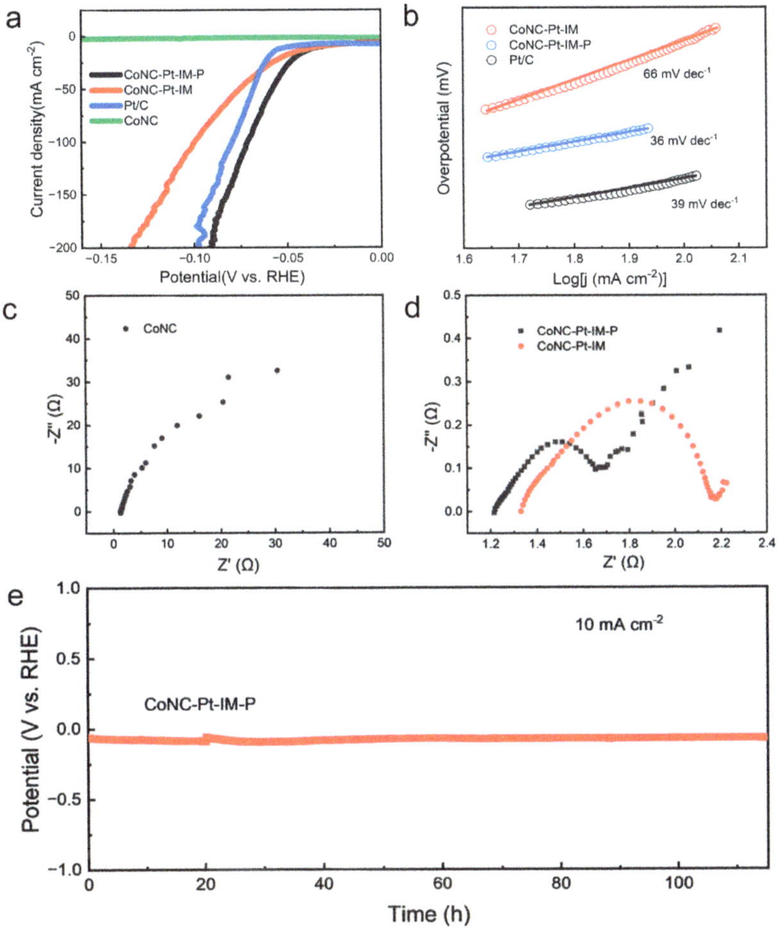

Figure 4. (**a**) LSV curves at 5 mV s^{-1} in 0.5 M H_2SO_4 of CoNC-Pt-IM-P, CoNC-Pt-IM, Pt/C, and CoNC. (**b**) Tafel slopes of CoNC-Pt-IM-P, CoNC-Pt-IM, and Pt/C. (**c**) EIS measurement of CoNC. (**d**) EIS measurement of CoNC-Pt-IM-P and CoNC-Pt-IM. (**e**) Stability test of CoNC-Pt-IM-P at current density of -10 mA cm^{-2} in 0.5 M H_2SO_4.

4. Conclusions

In this study, we successfully synthesized a novel CoNC-Pt-IM-P electrocatalyst through a combination of MOF template calcination and plasma treatment. Both CoNC-Pt-IM and CoNC-Pt-IM-P could retain their dodecahedral morphology, and elemental mapping confirmed the uniform surface loading of Pt. Moreover, the argon plasma treatment led to the noticeable aggregation of Co and Pt atoms so as to alter the electronic environment and enhance the intrinsic HER catalytic activity. CoNC-Pt-IM-P exhibited excellent HER activity, with an overpotential of only 31 mV at a current density of -10 mA cm^{-2}, which was lower than that of CoNC-Pt-IM (33 mV) and commercial Pt/C catalysts (50 mV). Meanwhile, its mass activity reached 4.9 A mg^{-1} Pt, which was 1.5 times that of CoNC-Pt-IM and 12.2 times that of the commercial Pt/C catalyst. Furthermore, CoNC-Pt-IM-P demonstrated remarkable stability, maintaining performance with negligible voltage changes over 110 h at a current density of -10 mA cm^{-2}. This work successfully used plasma treatment to achieve the controllable aggregation of Co and Pt atoms to enhance their catalytic activity, providing significant insights for the modification of multi-metal catalysts.

Supplementary Materials: The following supporting information can be downloaded at https://www.mdpi.com/article/10.3390/coatings14121569/s1. Figure S1. (a,b) SEM image of ZIF-67. (c,d) SEM image of CoNC. (e,f) SEM image of ZIF-67-Pt; Figure S2. HAADF-STEM–EDS mapping of the as-synthesized CoNC-Pt-IM; Figure S3. (a–c) Elemental distribution diagrams of CoNC-Pt-IM. (d–f) Elemental distribution di-agrams of CoNC-Pt-IM-P; Figure S4. (a) XRD patterns of ZIF-67 and ZIF-67-Pt. (b) Local XRD spectra of CoNC-Pt-IM-P; Figure S5. (a–c) XPS survey spectra of CoNC, CoNC-Pt-IM, and ZIF-67-Pt. (d) Atomic ratios of Pt and Co contents on catalyst surface by XPS peak area fitting. (e,f) High-resolution scan of Pt 4f XPS spectra. (g–i) Co 2p XPS spectra of CoNC, CoNC-Pt-IM, and ZIF-67-Pt; Figure S6. C1s XPS spectra of CoNC, CoNC-Pt-IM, and ZIF-67-Pt. (d–f) N1s XPS spectra of CoNC, CoNC-Pt-IM, and ZIF-67-Pt; Figure S7. LSV curves of CoNC-Pt-IM-P, CoNC-Pt-IM, Pt/C, CoNC, and ZIF-67-Pt; Figure S8. Tafel slopes of CoNC-Pt-IM-P, CoNC-Pt-IM, and CoNC; Figure S9. Cyclic voltammograms in 0.5 M of H2SO4 at scan of 20–100 mV s^{-1} in non-faradaic region for (a,b) CoNC, (c,d) CoNC-Pt-IM, and (e,f) CoNC-Pt-IM-Pt, respectively; Table S1. A summary of the deconvoluted parameters of Pt 4f XPS spectra in different catalysts; Table S2. A summary of the deconvoluted parameters of Co 2p XPS spectra in different catalysts.

Author Contributions: Conceptualization, G.Z.; Data Curation, J.L. and L.L.; Formal Analysis, Y.W.; Funding Acquisition, L.Z.; Investigation, G.Z., J.L. and L.L.; Project Administration, L.Z.; Supervision, L.Z.; Validation, Y.W.; Visualization, J.L.; Writing—Original Draft, G.Z.; Writing—Review and Editing, Y.W., L.L. and L.Z. All authors have read and agreed to the published version of the manuscript.

Funding: This research was funded by National Natural Science Foundation of China, and grant number include 22378119, 22075076, and 22208092.

Institutional Review Board Statement: Not applicable.

Informed Consent Statement: Not applicable.

Data Availability Statement: The original contributions presented in this study are included in the article/Supplementary Materials; further inquiries can be directed to the corresponding author.

Conflicts of Interest: The authors declare no competing interests.

References

1. Kou, Z.; Liu, Y.; Cui, W.; Yang, B.; Li, Z.; Rodriguez, R.D.; Zhang, Q.; Dong, C.L.; Sang, X.; Lei, L.; et al. Electronic structure optimization of metal–phthalocyanine via confining atomic Ru for all-pH hydrogen evolution. *Energy Environ. Sci.* **2024**, *17*, 1540–1548. [CrossRef]
2. Luo, W.; Yu, Y.; Wu, Y.; Ma, Z.; Ma, X.; Jiang, Y.; Shen, W.; He, R.; Su, W.; Li, M. Realizing efficient oxygen evolution at low overpotential via dopant-induced interfacial coupling enhancement effect. *Appl. Catal. B Environ.* **2023**, *336*, 122928. [CrossRef]
3. Rao, R.R.; Kolb, M.J.; Giordano, L.; Pedersen, A.F.; Katayama, Y.; Hwang, J.; Mehta, A.; You, H.; Lunger, J.R.; Zhou, H.; et al. Operando identification of site-dependent water oxidation activity on ruthenium dioxide single-crystal surfaces. *Nat. Catal.* **2020**, *3*, 516–525. [CrossRef]
4. Küspert, S.; Campbell, I.E.; Zeng, Z.; Balaghi, S.E.; Ortlieb, N.; Thomann, R.; Knäbbeler-Buß, M.; Allen, C.S.; Mohney, S.E.; Fischer, A. Ultrasmall and Highly Dispersed Pt Entities Deposited on Mesoporous N-doped Carbon Nanospheres by Pulsed CVD for Improved HER. *Small* **2024**, *20*, 2311260. [CrossRef] [PubMed]
5. Paudel, D.R.; Pan, U.N.; Ghising, R.B.; Dhakal, P.P.; Dinh, V.A.; Wang, H.; Kim, N.H.; Lee, J.H. Interface modulation induced by the 1T Co-WS_2 shell nanosheet layer at the metallic $NiTe_2$/Ni core–nanoskeleton: Glib electrode-kinetics for HER, OER, and ORR. *Nano Energy* **2022**, *102*, 107712. [CrossRef]
6. Fernández-Vidal, J.; Koper, M.T.M. Effect of a Physisorbed Tetrabutylammonium Cation Film on Alkaline Hydrogen Evolution Reaction on Pt Single-Crystal Electrodes. *ACS Catal.* **2024**, *14*, 8130–8137. [CrossRef] [PubMed]
7. Xu, W.; Zhang, D.; Wang, T.; Lai, J.; Wang, L. Non-platinum-based electrocatalysts for high performance acidic hydrogen evolution reactions in proton exchange membrane water electrolysis. *Appl. Catal. B Environ. Energy* **2025**, *361*, 124626. [CrossRef]
8. Reier, T.; Nong, H.N.; Teschner, D.; Schlögl, R.; Strasser, P. Electrocatalytic Oxygen Evolution Reaction in Acidic Environments-Reaction Mechanisms and Catalysts. *Adv. Energy Mater.* **2017**, *7*, 1601275. [CrossRef]
9. Zeng, L.; Zhao, Z.; Huang, Q.; Zhou, C.; Chen, W.; Wang, K.; Li, M.; Lin, F.; Luo, H.; Gu, Y.; et al. Single-Atom Cr–N_4 Sites with High Oxophilicity Interfaced with Pt Atomic Clusters for Practical Alkaline Hydrogen Evolution Catalysis. *J. Am. Chem. Soc.* **2023**, *145*, 21432–21441. [CrossRef]
10. Liu, Y.; Wang, Q.; Zhang, J.; Ding, J.; Cheng, Y.; Wang, T.; Li, J.; Hu, F.; Yang, H.B.; Liu, B. Recent Advances in Carbon-Supported Noble-Metal Electrocatalysts for Hydrogen Evolution Reaction: Syntheses, Structures, and Properties. *Adv. Energy Mater.* **2022**, *12*, 2200928. [CrossRef]

11. Gao, F.; Zhang, Y.; Wu, Z.; You, H.; Du, Y. Universal strategies to multi-dimensional noble-metal-based catalysts for electrocatalysis. *Coord. Chem. Rev.* **2021**, *436*, 213825. [CrossRef]
12. Yu, T.; Liu, G.; Nie, T.; Wu, Z.; Song, Z.; Sun, X.; Song, Y.F. Pt-Loaded CoFe-Layered Double Hydroxides for Simultaneously Driving HER and HzOR. *ACS Catal.* **2024**, *14*, 14937–14946. [CrossRef]
13. Zhang, H.; Zhao, M.; Liu, H.; Shi, S.; Wang, Z.; Zhang, B.; Song, L.; Shang, J.; Yang, Y.; Ma, C.; et al. Ultrastable FeCo Bifunctional Electrocatalyst on Se-Doped CNTs for Liquid and Flexible All-Solid-State Rechargeable Zn–Air Batteries. *Nano Lett.* **2021**, *21*, 2255–2264. [CrossRef]
14. Wang, C.; Guo, W.; Chen, T.; Lu, W.; Song, Z.; Yan, C.; Feng, Y.; Gao, F.; Zhang, X.; Rao, Y.; et al. Advanced noble-metal/transition-metal/metal-free electrocatalysts for hydrogen evolution reaction in water-electrolysis for hydrogen production. *Coord. Chem. Rev.* **2024**, *514*, 215899. [CrossRef]
15. Chen, W.; Zhu, X.; Wei, W.; Chen, H.; Dong, T.; Wang, R.; Liu, M.; Ostrikov, K.; Peng, P.; Zang, S.Q. Neighboring Platinum Atomic Sites Activate Platinum–Cobalt Nanoclusters as High-Performance ORR/OER/HER Electrocatalysts. *Small* **2023**, *19*, 2304294. [CrossRef]
16. Zhang, S.L.; Lu, X.F.; Wu, Z.-P.; Luan, D.; Lou, X.W. Engineering Platinum–Cobalt Nano-alloys in Porous Nitrogen-Doped Carbon Nanotubes for Highly Efficient Electrocatalytic Hydrogen Evolution. *Angew. Chem. Int. Ed.* **2021**, *60*, 19068–19073. [CrossRef] [PubMed]
17. Xiao, L.; Wang, Z.; Guan, J. 2D MOFs and their derivatives for electrocatalytic applications: Recent advances and new challenges. *Coord. Chem. Rev.* **2022**, *472*, 214777. [CrossRef]
18. Yu, S.; Li, J.; Du, Y.; Wang, Y.; Zhang, Y.; Wu, Z. Sulfur-modified MOFs as efficient electrocatalysts for overall water splitting. *Coord. Chem. Rev.* **2024**, *520*, 216144. [CrossRef]
19. Li, D.; Zong, Z.; Tang, Z.; Liu, Z.; Chen, S.; Tian, Y.; Wang, X. Total Water Splitting Catalyzed by Co@Ir Core–Shell Nanoparticles Encapsulated in Nitrogen-Doped Porous Carbon Derived from Metal–Organic Frameworks. *ACS Sustain. Chem. Eng.* **2018**, *6*, 5105–5114. [CrossRef]
20. Yang, W.; Cheng, P.; Li, Z.; Lin, Y.; Li, M.; Zi, J.; Shi, H.; Li, G.; Lian, Z.; Li, H. Tuning the Cobalt–Platinum Alloy Regulating Single-Atom Platinum for Highly Efficient Hydrogen Evolution Reaction. *Adv. Funct. Mater.* **2022**, *32*, 2205920. [CrossRef]
21. Zheng, X.; Wang, Z.; Zhou, Q.; Wang, Q.; He, W.; Lu, S. Precision tuning of highly efficient Pt-based ternary alloys on nitrogen-doped multi-wall carbon nanotubes for methanol oxidation reaction. *J. Energy Chem.* **2024**, *88*, 242–251. [CrossRef]
22. Zhao, K.; Zhang, J.; Li, H.; Zhang, X.; Hua, Y.; Di, L. Boosting HER performance by using plasma prepared N-doped CNTs to support Pt nanoparticles. *Int. J. Hydrog. Energy* **2024**, *90*, 1271–1278. [CrossRef]
23. Liang, L.; Jin, H.; Zhou, H.; Liu, B.; Hu, C.; Chen, D.; Wang, Z.; Hu, Z.; Zhao, Y.; Li, H.-W.; et al. Cobalt single atom site isolated Pt nanoparticles for efficient ORR and HER in acid media. *Nano Energy* **2021**, *88*, 106221. [CrossRef]
24. Mo, R.; Zhang, X.; Chen, Z.; Huang, S.; Li, Y.; Liang, L.; Tian, Z.Q.; Shen, P.K. Highly efficient PtCo nanoparticles on Co–N–C nanorods with hierarchical pore structure for oxygen reduction reaction. *Int. J. Hydrog. Energy* **2021**, *46*, 15991–16002. [CrossRef]
25. Hu, G.; Shang, L.; Sheng, T.; Chen, Y.; Wang, L. PtCo@NCs with Short Heteroatom Active Site Distance for Enhanced Catalytic Properties. *Adv. Funct. Mater.* **2020**, *30*, 2002281. [CrossRef]
26. Wang, Z.; Zhu, C.; Tan, H.; Liu, J.; Xu, L.; Zhang, Y.; Liu, Y.; Zou, X.; Liu, Z.; Lu, X. Understanding the Synergistic Effects of Cobalt Single Atoms and Small Nanoparticles: Enhancing Oxygen Reduction Reaction Catalytic Activity and Stability for Zinc-Air Batteries. *Adv. Funct. Mater.* **2021**, *31*, 2104735. [CrossRef]
27. Wang, Z.; Li, S.; Zhang, G.; Yu, X.; Zhao, Z.; Zhang, Y.; Shi, Y.; Zhu, H.B.; Xiao, X. Efficient electrocatalytic overall water splitting of porous 3D $CoPt_3$/a-FCWO-NS heterostructures: Morphology modulation and interfacial engineering for the formation of crystalline-amorphous nanosheets. *Appl. Catal. B Environ.* **2024**, *342*, 123387. [CrossRef]
28. Ao, X.; Zhang, W.; Zhao, B.; Ding, Y.; Nam, G.; Soule, L.; Abdelhafiz, A.; Wang, C.; Liu, M. Atomically dispersed Fe–N–C decorated with Pt-alloy core–shell nanoparticles for improved activity and durability towards oxygen reduction. *Energy Environ. Sci.* **2020**, *13*, 3032–3040. [CrossRef]
29. Chen, S.; Ma, L.; Wu, S.; Wang, S.; Li, Z.; Emmanuel, A.A.; Huqe, M.R.; Zhi, C.; Zapien, J.A. Uniform Virus-Like Co–N–Cs Electrocatalyst Derived from Prussian Blue Analog for Stretchable Fiber-Shaped Zn–Air Batteries. *Adv. Funct. Mater.* **2020**, *30*, 1908945. [CrossRef]
30. Duan, X.; Cao, F.; Ding, R.; Li, X.; Li, Q.; Aisha, R.; Zhang, S.; Hua, K.; Rui, Z.; Wu, Y.; et al. Cobalt-Doping Stabilized Active and Durable Sub-2 nm Pt Nanoclusters for Low-Pt-Loading PEMFC Cathode. *Adv. Energy Mater.* **2022**, *12*, 2103144. [CrossRef]
31. Yin, Y.; Liu, X.; Wei, X.; Li, Y.; Nie, X.; Yu, R.; Shui, J. Magnetically Aligned Co–C/MWCNTs Composite Derived from MWCNT-Interconnected Zeolitic Imidazolate Frameworks for a Lightweight and Highly Efficient Electromagnetic Wave Absorber. *ACS Appl. Mater. Interfaces* **2017**, *9*, 30850–30861. [CrossRef]
32. Liu, X.; Yan, W.; Song, J.; Song, H.; Chen, W.; Zhang, Y.; Chen, Y. Kirkendall effect induced the formation of hollow Co_2P in Co-N-C for ORR, OER, HER and flexible Zn-Air battery. *Chem. Eng. J.* **2024**, *492*, 152301. [CrossRef]
33. Sun, M.; Yun, S.; Dang, J.; Zhang, Y.; Liu, Z.; Qiao, D. 1D/3D rambutan-like Mott–Schottky porous carbon polyhedrons for efficient tri-iodide reduction and hydrogen evolution reaction. *Chem. Eng. J.* **2023**, *458*, 141301. [CrossRef]

34. Feng, J.; Wang, X.; Pan, H. In-situ Reconstruction of Catalyst in Electrocatalysis. *Adv. Mater.* **2024**, *36*, 2411688. [CrossRef]
35. Kuznetsova, I.; Lebedeva, O.; Kultin, D.; Perova, N.; Kalmykov, K.; Chernavskii, P.; Perov, N.; Kustov, L. Is a 2D Nanostructured Surface Capable of Changing the Corrosion and Magnetic Properties of an Amorphous Alloy? *Int. J. Mol. Sci.* **2023**, *24*, 13373. [CrossRef]

coatings

Article

First-Principles Study on the Electronic Structure and Optical Properties of $BiOIO_3$ Doped with As, Se, and Te

Fumei Lang [1], Xue Wen [1], Jibo Liu [2], Yineng Huang [1,3], Lili Zhang [1,3,*], Haiming Lu [1,4], Kaiye Jiang [1] and Baohua Zhang [1,5,*]

1 Xinjiang Laboratory of Phase Transitions and Microstructures in Condensed Matters, College of Physics and Technology, Yili Normal University, Yining 835000, China; langfm@sina.com (F.L.); wx19899036782@sina.com (X.W.); ynhuang@nju.edu.cn (Y.H.); haimlu@nju.edu.cn (H.L.); ylwlzll@163.com (K.J.)
2 School of Information Science and Engineering, Xinjiang University of Science and Technology, Korla 841000, China; liujibo1207@163.com
3 National Laboratory of Solid State Microstructures, School of Physics, Nanjing University, Nanjing 210093, China
4 Jiangsu Key Laboratory of Artificial Functional Materials, Department of Materials Science and Engineering, College of Engineering and Applied Sciences, Nanjing University, Nanjing 210093, China
5 Department of Physics, Changji University, Changji 831100, China
* Correspondence: suyi2046@sina.com (L.Z.); zbhcjxy@163.com (B.Z.)

Abstract

This study calculates the electronic structure and optical properties of intrinsic $BiOIO_3$ and X-$BiOIO_3$ (X = As, Se, or Te) using PBE (Perdew–Burke–Ernzerhof) and MBJ (Modified Becke–Johnson) functionals based on density functional theory, with MBJ showing better correlation with experimental values. The X-$BiOIO_3$ systems exhibit relative stability under MBJ potential and show crystal lattice distortion compared to intrinsic $BiOIO_3$, creating localized potential differences that enhance polarization and adjust the bandgap. Doping reduces the bandwidth and increases energy level density, promoting electron transitions. Consequently, based on the computational results presented in this paper, it can be inferred that both $BiOIO_3$ and X-$BiOIO_3$ facilitate water hydrolysis and oxygen generation due to their favorable energy band positions. Notably, Se-$BiOIO_3$ exhibits the highest visible light absorption capacity, which may enhance photocatalytic efficiency by strengthening the built-in electric field and promoting charge carrier generation.

Keywords: first principles; optical properties; electronic structure; non-metallic oxide; $BiOIO_3$

1. Introduction

Under specific lighting conditions, TiO_2 exhibits photocatalytic properties that enable water splitting for the production of clean and sustainable hydrogen energy, thereby driving traditional photocatalysis research in the scientific community [1]. Subsequent studies have revealed that conventional photocatalysts, such as TiO_2, ZnO, and SiO_2, possess the capability to not only facilitate water decomposition but also effectively degrade various pollutants [2]. However, the pronounced recombination rate of photo-generated electron–hole pairs in conventional photocatalytic materials significantly hampers their practical implementation in industrial production [3]. In recent years, the quest for novel photocatalytic materials [4] has emerged as a prominent trend in scientific research, garnering significant attention from numerous scholars.

As early as 2004, Grosso et al. [5] proposed the concept of utilizing ferroelectrics in the realm of photocatalysis, subsequently leading to the identification of several ferroelectric materials exhibiting exceptional photocatalytic activity [6,7]. Bismuth-based material $BiOIO_3$, as a representative ferroelectric catalyst, achieved removal efficiencies of 100%, 85.2%, and 65.2% for methyl orange, rhodamine B, and methylene blue, respectively, after 16 min of ultraviolet light exposure. Compared to commercial TiO_2 and other bismuth oxides, $BiOIO_3$ nanosheets exhibited significantly enhanced photocatalytic activity, with the apparent reaction rate constant for methyl orange being 10.26 times higher than that of TiO_2 [8]. The exceptional photocatalytic ability of $BiOIO_3$ can be attributed to its unique layered structure [9], comprising nested $[Bi_2O_2]^{2+}$ and $[IO_3]^-$ units. Notably, the latter exhibits a triangular pyramidal configuration, as illustrated in Figure 1. Due to the asymmetric distribution of its atoms, the separation of positive and negative charge centers gives rise to a dipole moment in the $[IO_3]^-$ triangular pyramidal system, with a flow of negative charge to positive charge. The polarization directions of $[IO_3]^-$ groups are systematically aligned along the C-axis, resulting in a pronounced macroscopic polarization effect [9] in the $BiOIO_3$, as well as the establishment of an inherent electric field due to spontaneous polarization. The intrinsic electric field [10] plays a pivotal role in facilitating rapid carrier migration and suppressing electron–hole recombination, thereby significantly enhancing the photocatalytic efficiency of $BiOIO_3$ [11–13]. As a result of these unique properties, $BiOIO_3$ holds great potential for broad applications in the field of photocatalysis.

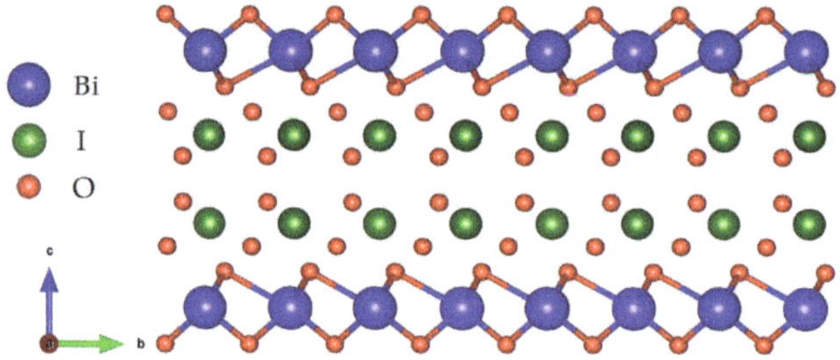

Figure 1. Model diagram of $BiOIO_3$.

Although the presence of an internal electric field in $BiOIO_3$ enhances its photocatalytic performance [11], its larger band gap width of 3.3 eV hinders further improvement in its photocatalytic ability [14]. Existing research has demonstrated that doping modification could enhance the photocatalytic performance of $BiOIO_3$. In their experimental studies, Huang et al. [15] employed $BiOIO_3$ and N-$BiOIO_3$ as photocatalysts for the degradation of various pollutants including Rhodamine B, bisphenol A, and tetracycline. Their findings revealed that N-$BiOIO_3$ demonstrated superior photocatalytic degradation efficiency compared to intrinsic $BiOIO_3$. Liu et al. [16] incorporated sulfur atoms into the oxygen sites of $BiOIO_3$, resulting in a downward shift of the conduction band edge and a reduction in the bandgap. Huang et al. [17] accomplished a visible-light photocatalytic performance in the layered bismuth-based photocatalyst $BiOIO_3$ through iodine ion doping. In contrast to pure $BiOIO_3$, the photoresponse range of iodine-doped $BiOIO_3$ was significantly expanded from ultraviolet light to visible light, while attaining a tunable bandgap. The aforementioned modifications result in a reduction in the energy required for electron transition and facilitate the process, thereby significantly enhancing its photocatalytic efficacy. The enhancement achieved through non-metal doping demonstrates the remarkable potential of this modification strategy to improve the photocatalytic performance of $BiOIO_3$. In order to evaluate the photocatalytic performance of $BiOIO_3$ substitutional doped non-metallic elements, this study employs first principles to calculate the electronic structure and optical properties of $BiOIO_3$-doped X. The rationale for selecting As, Se, and Te as doping elements lies in their classification within similar groups (Group VA and Group VIA) of the periodic table. These elements exhibit analogous chemical and physical properties, enabling them to form substitutional or interstitial dopants within the same host material matrix. Consequently, this facilitates the effective modification of the material's electronic properties. The incorporation of arsenic, selenium, and tellurium can modulate the energy band structure of materials, particularly semiconductors, thereby enhancing light absorption capabilities and improving the separation efficiency of photo-generated carriers, which ultimately boosts photocatalytic activity. Additionally, the introduction of As, Se, and Te atoms significantly alters the carrier concentration of the material, thereby influencing its charge transport properties and reactivity. The obtained calculation results will offer theoretical guidance for the synthesis of $BiOIO_3$ catalysts that are modified using the doping method.

2. Materials and Methods

In this study, the crystal structure of $BiOIO_3$ is orthorhombic, exhibiting a space group of $Pca2_1$ (No. 29) [11]. This non-centrosymmetric arrangement (NCS) consists of interleaved $[Bi_2O_2]^{2+}$ and $[IO_3]^-$ units. The model of doped $BiOIO_3$, as depicted in Figure 2, demonstrates the utilization of one X atom being substituted for one I atom. Hence, the doping concentration of X-$BiOIO_3$ is precisely 6.25%. The replaced position of atom I must be excluded outside the boundaries in order to mitigate boundary effects. Moreover, within the unit cell, any position of atom I is symmetrical and satisfies the requirements of symmetry. Hence, the replacement position of atom I can be chosen arbitrarily within the unit cell without impacting the final calculation results. In this study, we ultimately selected the coordinate position of replaced atom I as (0.748, 0.366, 0.335). The electronic configurations of the elements discussed in this article are as follows, Bi: $6s^2$ $6p^3$; I: $5s^2$ $5p^5$; O: $2s^2$ $2p^4$; As: $4s^2$ $4p^3$; Se: $4s^2$ $4p^4$; Te: $5s^2$ $5p^4$.

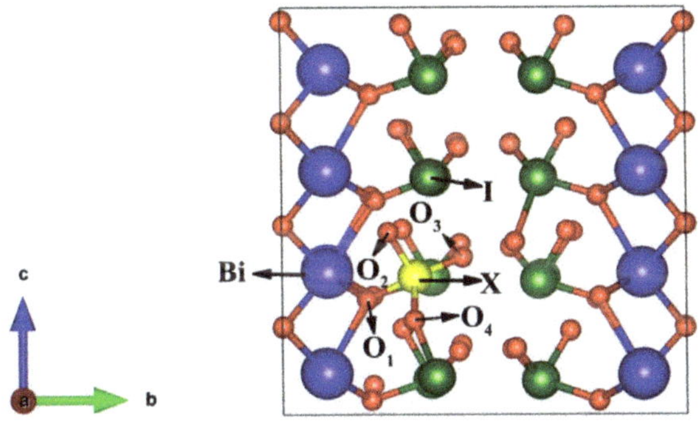

Figure 2. Model diagram of $BiOIO_3$ doped with non-metallic elements X. The symbols O_1, O_2, O_3, and O_4 represent oxygen atoms with different symmetries in the $BiOIO_3$ crystal lattice.

The calculations are predominantly performed using the Vienna Ab-initio Simulation Package (VASP 6) [18], specifically employing the generalized gradient approximation (GGA) within density functional theory (DFT) and utilizing the MBJ semi-local exchange-correlation potential [19] for material characterization. The cut-off energy for the $BiOIO_3$ was set at 550 eV. The K-point grid was determined according to Monkhorst's scheme [20], employing a $4 \times 4 \times 4$ configuration. The convergence criterion for internal stress was established to be no greater than 0.02 GPa, while the energy convergence threshold was defined as being within 2×10^{-6} eV/atom. Additionally, the self-consistency iteration convergence accuracy was specified to be 2×10^{-6} eV/atom, and the MBJ parameter c value was determined to be 1.34. Due to the discrepancy between results obtained from calculations using monoclinic cells and actual conditions, an expanded cell model was employed to optimize computational resources while ensuring accuracy. Ultimately, a $2 \times 2 \times 1$ $BiOIO_3$ supercell model was selected.

3. Results and Discussion

3.1. Geometric Optimization Results

To improve the precision of the computational results, this study conducts geometric optimization on the systems, with the outcomes presented in Table 1. After optimization, it is observed that the lattice constant ratio (c/a) for intrinsic $BiOIO_3$ is 1.013, exhibiting a negligible deviation of only 0.2% from the experimental value of 1.0159. This observation suggests that the computational parameters employed in this study exhibit reliability and generate precise outcomes. The volume of each doped system changes, indicating that these systems experience lattice distortion relative to the calculated values of intrinsic $BiOIO_3$. Considering the inherent non-centrosymmetric (NCS) nature of $BiOIO_3$, it naturally induces a polarization effect. The doping further enhances this phenomenon, facilitating the generation of localized potential differences within the systems and augmenting the migration rate of photo-generated electron–hole pairs.

Table 1. The lattice constant and volume of each model.

Model	a/nm	b/nm	c/nm	V/nm^3
$BiOIO_3$ (Experiment) [21]	0.5658	1.1039	0.5748	/
$BiOIO_3$ (This work)	0.5712	1.1263	0.5788	1.4898
As-$BiOIO_3$	0.5698	1.1311	0.5779	1.4899
Se-$BiOIO_3$	0.5706	1.1237	0.5780	1.4826
Te-$BiOIO_3$	0.5695	1.1272	0.5775	1.4830

3.2. Structural Stability of $BiOIO_3$ and X-$BiOIO_3$

To conduct a detailed investigation of the stability of the four systems, the formation energy and binding energy are calculated for each system. The binding energy [22] and formation energy [23] serve to characterize the stability of the system and the ease of doping, respectively. A lower formation energy indicates a greater ease of doping, while a lower binding energy correlates with the enhanced stability of the system. Additionally, the doping formation energy (E_f) [24] and binding energy (E_b) [25] for different systems are determined using Equations (1) and (2).

$$E_f = E_{XBIO} - E_{BIO} - m\mu_x + n\mu_I \quad (1)$$

$$E_b = 1/N(E_{tot} - E_{BIO}) \quad (2)$$

In the equation, E_{tot} is the total energy of all the atoms of the system in their free state. E_{XBIO} represents the total energy of the system after doping. E_{BIO} denotes the total energy of the $BiOIO_3$. μ_x and μ_I denote the chemical formulas of the dopant atom X and the host atom I, respectively. m and n denote the quantities of dopant atoms and substituted atoms, respectively (in this study, m = n = 1). N denotes the total number of atoms present in the system. All the calculated results are listed in Table 2. The binding energy and formation energy of all doped systems exhibit negative values, implying that the formation of these systems is straightforward. The As-doped $BiOIO_3$ demonstrates optimal stability and facile formation, as evidenced by its significantly lower binding energy and formation energy compared to the other systems observed.

Table 2. The total energy, binding energy, and formation energy of each model before and after doping.

Model	E/eV	E_b/eV	E_f/eV
$BiOIO_3$	−477.2118	/	/
As-$BiOIO_3$	−479.9295	−0.3186	−4.9405
Se-$BiOIO_3$	−479.3817	−0.3183	−2.1449
Te-$BiOIO_3$	−479.9295	−0.3185	−2.6956

E represents the total energy of $BiOIO_3$ and the X-$BiOIO_3$ systems.

3.3. Electronic Structure of $BiOIO_3$ and X-$BiOIO_3$

To further investigate the variations in electronic structure among different systems, this study computes the energy band and electron state density for each system. The high symmetry points in the Brillouin region of the $BiOIO_3$ cells are denoted by Γ (0, 0, 0), M (0.5, 0, 0), K (0.5, 0.5, 0), Z (0, 0.5, 0), and Γ (0, 0, 0), as shown in Figure 3. The line with zero energy is designated as the Fermi level. The density of states for intrinsic $BiOIO_3$ is presented in Figure 3a,b, calculated using the PBE functional [26] and MBJ functional, respectively. The results indicate that the bandgap value obtained from the PBE calculation is 2.043 eV, which deviates by approximately 38% from the experimental value (3.3 eV) [14]. This underestimation occurs because when solving the Cohen–Shen equation, PBE does not consider the excited states of the system, thereby leading to the underestimation of the band gap. To solve this problem, the kinetic energy density term is added to the MBJ function on the basis of GGA-PBE in order to improve the accuracy of energy and band structure calculations. Therefore, the MBJ potential achieves a significantly higher accuracy in energy and band structure calculations [27]. To enhance the accuracy of the bandgap, we employ the MBJ functional to perform a recalculation of the band structure of materials. The recalculated result of 3.308 eV exhibits a deviation of approximately 0.24% from the experimental value, thereby substantiating its reliability. Consequently, we will employ the MBJ functional approach for subsequent sections to conduct further investigations into the electronic structure.

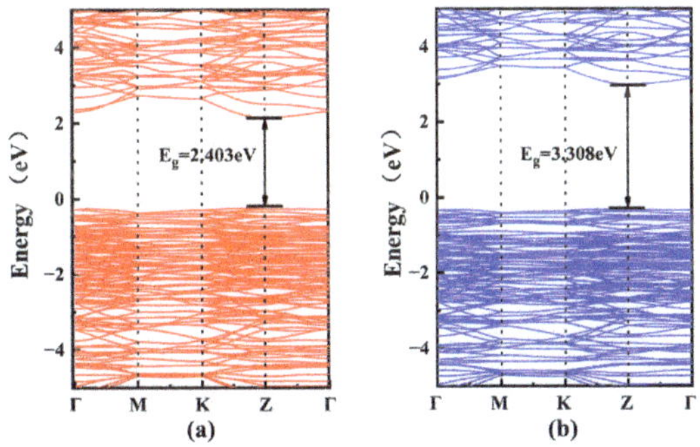

Figure 3. $BiOIO_3$ band diagrams. (**a**) $BiOIO_3$-PBE; (**b**) $BiOIO_3$-MBJ.

The band structure of non-metallic doped materials is depicted in Figure 4. From the figure, it can be observed that the bandgap values of As-$BiOIO_3$, Se-$BiOIO_3$, and Te-$BiOIO_3$ are calculated as 2.590 eV, 3.171 eV, and 2.867 eV, respectively. These values are found to be smaller than the intrinsic bandgap of $BiOIO_3$ by 21.7%, 4.1%, and 13.3%, respectively, facilitating an easier transition of valence band electrons to the conduction band. Consequently, this enhances the probability of electron transition and further modulates the optical performance of the systems. Among them, As-$BiOIO_3$ exhibits the smallest bandgap, accompanied by an impurity level at 2.072 eV that facilitates electron transitions and enhances the probability of such transitions. Accordingly, doping leads to a denser energy band structure, thereby promoting electron migration.

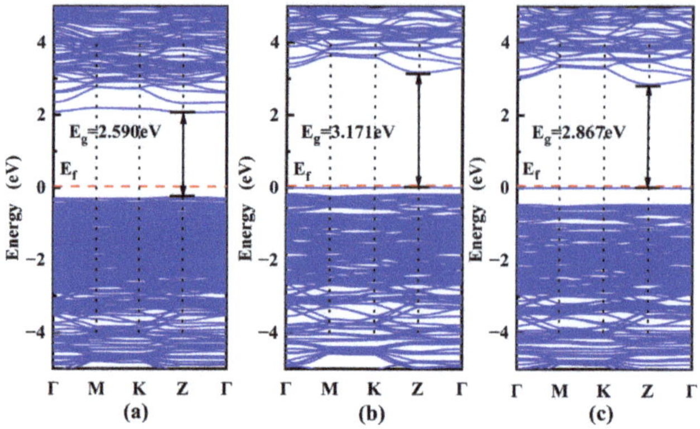

Figure 4. Band diagrams of doping systems. (**a**) As-$BiOIO_3$; (**b**) Se-$BiOIO_3$; (**c**) Te-$BiOIO_3$.

To further investigate the impact of doping on the electronic structure of the system, we present herein the total and partial density of states prior to and subsequent to doping, as visually depicted in Figure 5. Figure 5a presents the electron density of states for the intrinsic $BiOIO_3$. The valence band maximum (VBM) primarily originates from the 2p orbitals of oxygen and the 6s orbitals of Bi, while the conduction band minimum (CBM) predominantly comprises the 5p orbitals of I, along with the 6p orbitals of Bi and the 2p orbitals of O. Figure 5b presents the electron density of states for the As-doped $BiOIO_3$. The VBM primarily originates from the O-2p, Bi-6s, I-5p, and I-5s orbitals, with a minor contribution from the As-4p orbital. In contrast, the CBM is predominantly composed of the I-5p, O-2s, Bi-6p, and O-2p orbitals, as well as including contributions from the As-4s and As-4p orbitals. Furthermore, the introduction of an impurity element leads to the emergence of a defect energy level, thereby contributing to a reduction in the bandgap width within the system. The presence of this impurity level facilitates electron transitions by serving as a "bridge", thereby augmenting the likelihood of such transitions and enhancing the optical performance of the system. Figure 5c illustrates the electron density of states for Se-doped $BiOIO_3$, demonstrating that the contributions of Bi, O, and I to the system are largely consistent with those in intrinsic $BiOIO_3$. However, the incorporation of Se leads to an upward shift in the valence band, causing the Fermi level to traverse through it and exhibit typical characteristics of a p-type semiconductor. Additionally, this upward shift of the valence band also leads to a slight reduction in the bandgap. The electron density of states for Te-doped $BiOIO_3$, as shown in Figure 5d, exhibits a striking resemblance to that of Se-doped $BiOIO_3$. The incorporation of Te induces an upward shift in the valence band, thereby conferring p-type semiconductor characteristics upon the system. The conduction band also undergoes a downward shift, resulting in a significant reduction in the width of the bandgap due to the influence exerted by the Te-5s orbital. Consequently, the introduction of As, Se, and Te as dopants in $BiOIO_3$ leads to a reduction in the energy required for electron transitions, thereby indicating an augmentation in the optical performance of the systems.

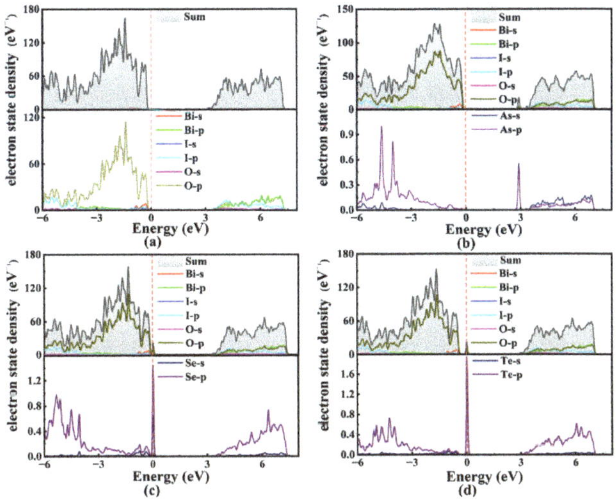

Figure 5. Total electron state density and partial electron state density. (**a**) Intrinsic $BiOIO_3$; (**b**) As-$BiOIO_3$; (**c**) Se-$BiOIO_3$; (**d**) Te-$BiOIO_3$.

3.4. The Two-Dimensional Electron Density and Bader Charges of $BiOIO_3$ and X-$BiOIO_3$

The electron density diagrams on the (100) plane of $BiOIO_3$ before and after doping are presented in Figure 6. It is evident that doping induces alterations in the distribution of electron clouds within the systems, implying a modification in atomic bonding. The introduction of an As atom into $BiOIO_3$ leads to a shift in the bonding between the As and O atoms due to the deviation in atomic positions, resulting in enhanced charge delocalization near the As atoms, as depicted in Figure 6b. Meanwhile, no significant alteration in the electron cloud distribution is observed for the Se-$BiOIO_3$ and Te-$BiOIO_3$ systems, as depicted in Figure 6c,d. However, when compared to intrinsic $BiOIO_3$, the Se atom demonstrates a higher degree of delocalization while enhancing the localization of the adjacent O atom. This suggests a transfer of electrons from the vicinity of the Se atom to the O atom. When one Te atom replaces one I atom, the degree of electron cloud overlap with the neighboring O atoms decreases, indicating a weakening in the bonding interaction between Te and its adjacent O atoms. Consequently, upon the substitution of I atoms by X atoms, the electron cloud undergoes varying degrees of changes, indicating that doping modifies the internal electron arrangement of the system and facilitates electron transfer [22]. The observed phenomenon can potentially be attributed to the disparity in non-metallic properties among the four atoms. In accordance with Pauling electronegativity, iodine (I) exhibits a higher electronegativity value of 2.66 compared to arsenic (As) at 2.18, selenium (Se) at 2.55, and tellurium (Te) at 2.1. Consequently, the non-metallic characteristics of these atoms follow the order I > Se > As > Te. Thus, upon the substitution of iodine atoms with X atoms, there is a reduction in electron cloud density near this position.

To further investigate the charge transfer dynamics in each system following doping, we employed the Bader charge method to analyze the charges of the oxygen atoms adjacent to the doping sites. The obtained results are presented in Table 3, with atomic numbers cross-referenced in Figure 1.

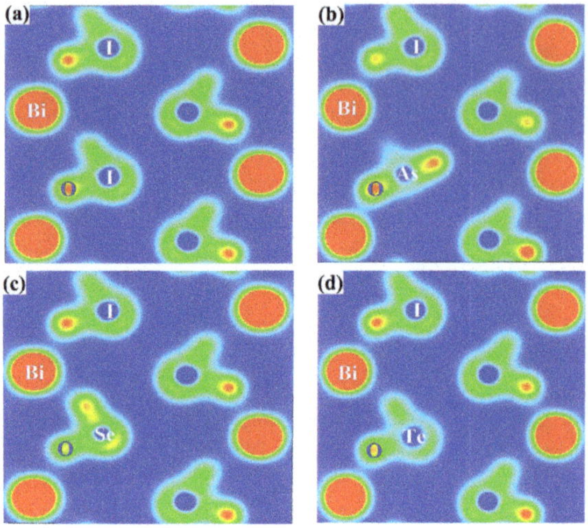

Figure 6. Two-dimensional electron density maps of plane (100). (**a**) Intrinsic $BiOIO_3$; (**b**) As-$BiOIO_3$; (**c**) Se-$BiOIO_3$; (**d**) Te-$BiOIO_3$. The red area represents the region with a higher electron density, while the blue area represents the region with a lower electron density. Yellow region represents a positive charge area.

Table 3. Bader charge of O atom near the doping site.

Model	The Charge of O Atoms				
	O_1	O_2	O_3	O_4	Average
$BiOIO_3$	6.8260	6.9150	6.9662	6.9119	6.9048
As-$BiOIO_3$	7.0762	7.0254	6.9087	7.1356	7.0364
Se-$BiOIO_3$	7.0203	7.0499	6.9736	7.0125	7.0141
Te-$BiOIO_3$	7.0304	7.1611	7.0102	7.1411	7.0857

The table demonstrates a high degree of consistency in the charges of the four nearest oxygen atoms surrounding the I atom in intrinsic $BiOIO_3$, implying an even distribution of electrons among these oxygen atoms and a lack of significant distortion. Upon doping with the non-metal X, a pronounced distortion in the charge distribution of oxygen is observed. In comparison to the intrinsic $BiOIO_3$, the charges of the O atoms adjacent to the doping sites in the doped system are all elevated due to the higher electronegativity of I compared to that of the X atoms. This observation is consistent with electron density analysis findings. The table further demonstrates that variations in the doping element exert a significant influence on the dynamics of charge transfer. In As-$BiOIO_3$, electron transfer predominantly occurs towards the O_4 direction, whereas in Se-$BiOIO_3$, there is a relatively higher electron flux towards the O_2 direction; similarly, in Te-$BiOIO_3$, there is also a pronounced movement of electrons towards the O_2 direction. The charge transfer direction in the doped system has shifted compared to intrinsic $BiOIO_3$, owing to an increased degree of distortion induced by doping. This alteration in the internal charge distribution aligns with previous analytical findings. Hence, it is evident that non-metal X doping can effectively modify the electronic structure and enhance the distortion degree of the system, resulting in a greater deviation between positive and negative charges. Consequently, this strengthens the polarization field and modulates its optical properties [28].

3.5. Optical Properties of $BiOIO_3$ and X-$BiOIO_3$

3.5.1. Alignment with Band Structure Edges

The photocatalyst is evaluated in accordance with the relative position of its semiconductor conduction band maximum (CBM) and valence band (VBM) values in relation to the standard hydrogen electrode minimum (NHE). The determination of the band edge potential on the NHE scale typically relies on Equations (3) and (4) [29]. In both equations, X represents the absolute electronegativity of the material. The energy level of the free electron on the NHE scale, approximately 4.5 eV, is denoted as E_{elec}. Instead, E_g refers to the band gap of a semiconductor.

$$E_{VBM} = X - E_{elec} + 0.5E_g \quad (3)$$

$$E_{CBM} = E_{VBM} - E_g \quad (4)$$

The valence band edge potential of $BiOIO_3$ and X-$BiOIO_3$ at the NHE scale is depicted in Figure 7. The CBM of $BiOIO_3$ and X-$BiOIO_3$ lies above the H^+/H_2 potential, while the VBM is beneath the O_2/H_2O potential, suggesting that $BiOIO_3$ and X-$BiOIO_3$ possess the capability of water hydrolysis and oxygen generation. Nevertheless, due to the significant reduction in the bandgap width when Se is doped into $BiOIO_3$, the energy required for electron transition becomes smaller. Hence, it can be inferred that Se-$BiOIO_3$ has a superior water splitting photocatalytic activity among the four systems.

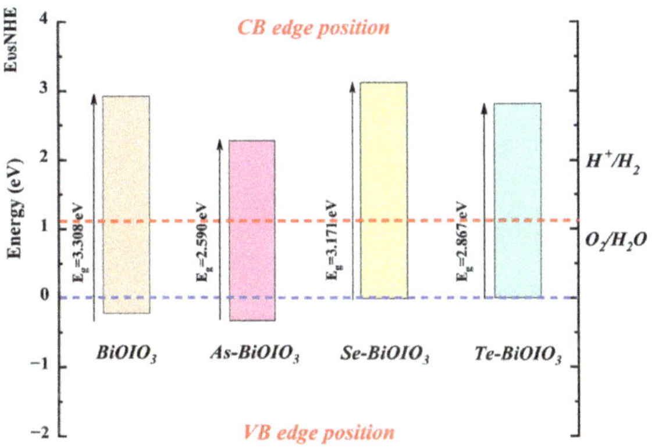

Figure 7. The potentials of $BiOIO_3$ and X-$BiOIO_3$ with edge characteristics. In the diagram, the blue dashed line represents the energy level of H^+/H_2, while the red dashed line corresponds to the energy level of O_2/H_2O.

3.5.2. Absorption Spectrum

Doping significantly alters the optical properties of a crystal by introducing impurities that modify the electronic states within the system, thereby enhancing its light responsiveness and influencing its optical characteristics. In this study, we analyze optical absorption graphs and dielectric diagrams to investigate the changes in the optical characteristics of the doped systems.

The optical absorption coefficient $\alpha(\omega)$ of a crystal can be determined by utilizing the real part $\varepsilon_r(\omega)$ and the imaginary part $\varepsilon_i(\omega)$ of the dielectric function, as demonstrated in Equations (5) and (6) [30].

$$\varepsilon(\omega) = \varepsilon_r(\omega) + i\varepsilon_i(\omega) \quad (5)$$

$$\alpha(\omega) = \sqrt{2}(\omega)\left[\sqrt{\varepsilon_r^2(\omega) + \varepsilon_i^2(\omega)} - \varepsilon_r(\omega)\right]^{\frac{1}{2}} \quad (6)$$

Theoretically, the optical absorption coefficient of X-$BiOIO_3$ systems exhibits a remarkable enhancement within the visible light range, as illustrated in Figure 8. In comparison to intrinsic $BiOIO_3$, X-$BiOIO_3$ systems demonstrate a significant red shift towards lower energy levels, indicating an augmented light-responsive capacity. These findings suggest that the introduction of As, Se, and Te can significantly enhance the optical absorption capacity of this system. The comparative analysis reveals that among all doping systems, Se-$BiOIO_3$ exhibits the highest light absorption capacity, thereby indicating its superior photocatalytic potential.

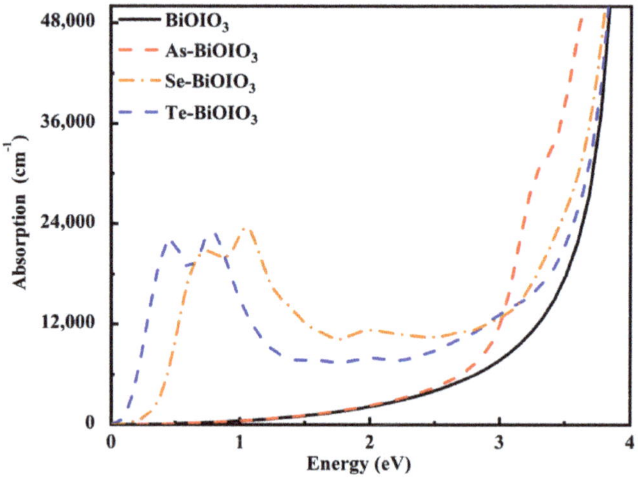

Figure 8. The absorption spectrum diagram of $BiOIO_3$ and X-$BiOIO_3$ systems.

3.5.3. Dielectric Function

The dielectric constant characterizes the response of materials to external energy. Figure 9a illustrates the calculated real part of the dielectric function for the four systems both before and after doping, highlighting the trend in the dielectric constant as it varies with incident light energy. The greater the dielectric constant, the stronger the system's polarization capability and internal electric field strength. When the incident light energy approaches zero, the corresponding value on the vertical axis represents the static dielectric constant. In the figure, the static dielectric constant of the intrinsic $BiOIO_3$ is calculated at 4.929, whereas the static dielectric constants for the X-$BiOIO_3$ systems are recorded as 4.8798, 7.567, and 5.938, respectively. In comparison to intrinsic $BiOIO_3$, the dielectric constants of the doped systems exhibit notable variations. Specifically, the static dielectric constant of As-$BiOIO_3$ is lower than that of intrinsic $BiOIO_3$, suggesting that the incorporation of As diminishes the polarization capability of the system. However, this reduction amounts to only 1.20% relative to intrinsic $BiOIO_3$ and can be considered negligible. Additionally, the dielectric constants of Se-$BiOIO_3$ and Te-$BiOIO_3$ exhibit increases of 53.51% and 20.47%, respectively, compared to the intrinsic $BiOIO_3$. This suggests that doping with Se and Te alters the polarization capability of $BiOIO_3$, leading to an enhancement in internal electric field strength and the accelerated movement of charge carriers within the doped systems, thereby improving charge binding capacity, which aligns with the previous analysis.

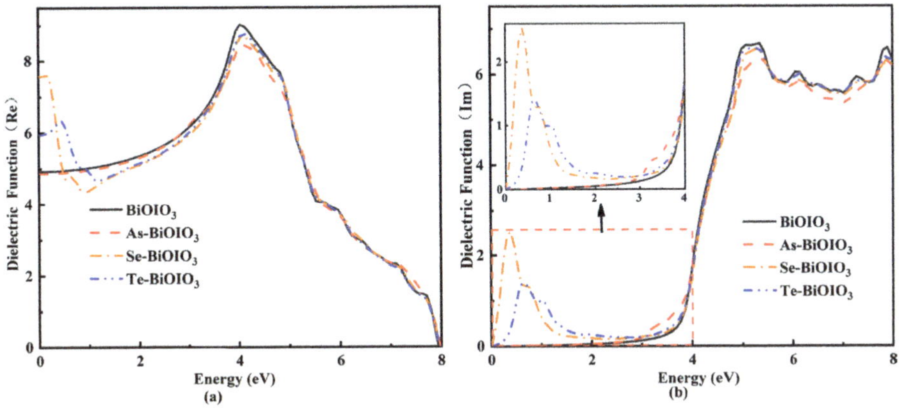

Figure 9. $BiOIO_3$ and X-$BiOIO_3$ dielectric function diagrams: (**a**) the dielectric real part; (**b**) the dielectric imaginary part.

Figure 9b illustrates the calculated imaginary part of the dielectric function before and after doping, revealing significant variations among the three X-$BiOIO_3$ systems within the 0–4 eV range. The imaginary part of the dielectric function represents the energy dissipated due to the induction of a large number of electric dipoles within the system in response to external energy stimulation. This value quantitatively reflects the degree of electronic transition, and its peak corresponds to the strengthened inter-band electronic transition of the system under the action of an external electric field. In addition, this value is also related to the probability of the system absorbing external energy [31]. Notably, Se-$BiOIO_3$ exhibits a small peak at 0.370 eV, reaching a maximum value of 2.5236 eV, attributed to an electron transition between the 4p state of Se and the 2p state of O. At 0.592 eV, a new peak is observed in the Te-$BiOIO_3$ system, attributed to the introduction of impurities that induce orbital hybridization and coupling at this energy level. In conclusion, the peak positions of the Se-$BiOIO_3$ and Te-$BiOIO_3$ systems following doping exhibit a tendency to shift towards the low-energy region. This observation aligns with the conclusions derived from band structure analysis and is also associated with the impurity levels introduced by doping. The doping of Se effectively reduces the bandgap of intrinsic $BiOIO_3$, enhances its light absorption coefficient, and strengthens the polarization electric field within the system. Consequently, it can be inferred that this system exhibits superior optical performance.

4. Conclusions

The energy band of intrinsic $BiOIO_3$ is calculated in this study using the PBE functional and MBJ functional based on the first principles of density functional theory. The results obtained under the MBJ functional exhibit superior agreement with experimental values. The X-$BiOIO_3$ systems demonstrate relative stability when calculated using the MBJ potential. In comparison to intrinsic $BiOIO_3$, the X-$BiOIO_3$ systems exhibit crystal lattice distortion, thereby facilitating the generation of localized potential differences within the system. The intensification of polarization promotes the adjustment of the bandgap. The bandwidth of X-$BiOIO_3$ reduces and the energy band distribution becomes denser. The diminished bandwidth and the denser energy levels of X-$BiOIO_3$ suggest that doping enhances the probability of electron transitions. The CBM of $BiOIO_3$ and X-$BiOIO_3$ lies above the H^+/H_2 potential, while the VBM is beneath the O_2/H_2O potential, suggesting that $BiOIO_3$ and X-$BiOIO_3$ possess the capability of water hydrolysis and oxygen generation. The expansion of the light absorption range significantly enhances the optical performance of the system. In contrast, the Se-$BiOIO_3$ system exhibits the highest absorption coefficient within the visible light region, followed by that of the Te-$BiOIO_3$ system. Furthermore, both Se and Te doping can modify the polarization capability of $BiOIO_3$. Based on the aforementioned conclusions, it can be deduced that doping with the non-metallic element Se enhances the internal electric field strength, accelerates charge carrier generation within the system, improves charge binding capability, increases electron mobility, and ultimately elevates the photocatalytic efficiency of $BiOIO_3$-based materials, thereby paving a new pathway for the development of photocatalytic materials.

Author Contributions: Conceptualization: F.L., J.L., Y.H., L.Z. and B.Z.; data curation: F.L. and L.Z.; formal analysis: F.L., X.W. and L.Z.; funding acquisition: F.L. and L.Z.; investigation: F.L. and X.W.; methodology: F.L., J.L., L.Z., H.L. and K.J.; project administration: L.Z.; supervision: L.Z.; validation: H.L. and B.Z.; visualization: X.W. and K.J.; writing—original draft: F.L. and J.L.; writing—review and editing: Y.H., L.Z. and B.Z. All authors have read and agreed to the published version of the manuscript.

Funding: This research was funded by the Open Project of Xinjiang Condensed Matter Phase Transformation and Microstructure Laboratory (grant number XJDX0912Z2404), the Science and Technology Plan Project of Yili Kazakh Autonomous Prefecture (grant number YZ2022B021), the Xinjiang Province Graduate Research and Innovation Project (grant number XJ2024G260), and the Innovation and Entrepreneurship Training Project of Yili Normal University Students (grant number S202310764004).

Institutional Review Board Statement: Not applicable.

Informed Consent Statement: Not applicable.

Data Availability Statement: The original contributions presented in the study are included in the article; further inquiries can be directed to the corresponding authors.

Acknowledgments: We gratefully acknowledge HZWTECH for providing computation facilities.

Conflicts of Interest: The authors declare no conflicts of interest.

References

1. Fujishima, A.; Honda, K. Electrochemical Photolysis of Water at a Semiconductor Electrode. *Nature* **1972**, *238*, 37–38. [CrossRef] [PubMed]
2. Chatterjee, D.; Dasgupta, S. Visible light induced photocatalytic degradation of organic pollutants. *J. Photochem. Photobiol. C Photochem. Rev.* **2005**, *6*, 186–205. [CrossRef]
3. Ma, S.; Yu, X.; Li, W.; Kong, J.; Long, D.; Bai, X. Bismuth-based photocatalysts for pollutant degradation and bacterial disinfection in sewage system: Classification, modification and mechanism. *Environ. Res.* **2024**, *264*, 120297. [CrossRef]
4. El-Bahy, S.M.; Arshad, J.; Munir, S.; Chaudhary, K.; Alhashmialameer, D.; Eddy, D.R.; Warsi, M.F.; Shahid, M. Improved photocatalytic performance of a new silver doped $BiSbO_4$ photocatalyst. *Ceram. Int.* **2022**, *48*, 23915. [CrossRef]
5. Grosso, D.; Boissière, C.; Smarsly, B.; Brezesinski, T.; Pinna, N.; Albouy, P.A.; Amenitsch, H.; Antonietti, M.; Sanchez, C. Periodically ordered nanoscale islands and mesoporous films composed of nanocrystalline multimetallic oxides. *Nat. Mater.* **2004**, *3*, 787–792. [CrossRef] [PubMed]
6. Mohan, S.; Subramanian, B. A strategy to fabricate bismuth ferrite ($BiFeO_3$) nanotubes from electrospun nanofibers and their Solar light-driven photocatalytic properties. *RSC Adv.* **2013**, *3*, 23737–23744. [CrossRef]
7. Cui, D.H.; Zheng, Y.F.; Song, X.C. Hydrothermal synthesis, characterisation and photocatalytic properties of $BiOIO_3$ nanoplatelets. *J. Exp. Nanosci.* **2016**, *11*, 1000–1010. [CrossRef]
8. Li, J.; Xie, J.; Zhang, X.; Lu, E.; Cao, Y. The Solid-State Synthesis of $BiOIO_3$ Nanoplates with Boosted Photocatalytic Degradation Ability for Organic Contaminants. *Molecules* **2023**, *28*, 3681. [CrossRef] [PubMed]
9. Zhi, Z.; Chen, T.Z.; Shui, J.Y.; Zhi, Y.Y.; Yun, Y. Ferroelectric polarization effect promoting the bulk charge separation for enhance the efficiency of photocatalytic degradation. *Chem. Eng. J.* **2021**, *410*, 128430.
10. Huang, H.; Chen, F.; Reshak, A.H.; Auluck, S.; Zhang, Y. Insight into crystal-structure dependent charge separation and photo-redox catalysis: A combined experimental and theoretical study on $Bi(IO_3)_3$ and $BiOIO_3$. *Appl. Surf. Sci.* **2018**, *458*, 129–138. [CrossRef]

11. Dong, X.D.; Yao, G.Y.; Liu, Q.L.; Zhao, Q.M.; Zhao, Z.Y. Spontaneous polarization effect and photocatalytic activity of layered compound of $BiOIO_3$. *Inorg. Chem.* **2019**, *58*, 2–9. [CrossRef] [PubMed]
12. Patiphatpanya, P.; Phuruangrat, A.; Thongtem, S.; Kungwankunakorn, S.; Thongtem, T. Effect of microwave power on phase, morphology, and photocatalytic properties of $BiOIO_3$ nanostructure. *J. Aust. Ceram. Soc.* **2019**, *55*, 501–506. [CrossRef]
13. Sun, Y.; Xiong, T.; Dong, F.; Huang, H.; Cen, W. Interlayer-I-doped $BiOIO_3$ nanoplates with an optimized electronic structure for efficient visible light photocatalysis. *Chem. Commun.* **2016**, *52*, 8243–8246. [CrossRef]
14. Nguyen, S.D.; Yeon, J.; Kim, S.; Halasyamani, P.S. $BiO(IO_3)$: A New Polar Iodate that Exhibits an Aurivillius-Type $(Bi_2O_2)^{2+}$ Layer and a Large Shg Response. *J. Am. Chem. Soc.* **2011**, *133*, 12422–12425. [CrossRef] [PubMed]
15. Huang, L.; Wang, Y.; Li, Y.; Huang, S.; Xu, Y.; Xu, H.; Li, H. Calcination synthesis of N-doped $BiOIO_3$ with high Led-light-driven photocatalytic activity. *Mater. Lett.* **2019**, *246*, 219–222. [CrossRef]
16. Liu, Y.L.; Zhu, Z.R.; Liu, Y.Q.; Jiang, W.; Yang, L.; Zi, J.X.; Shuai, Q.; Yu, B.W.Y.; Meng, J.B. First principles insight on enhanced photocatalytic performance of sulfur-doped bismuth oxide iodate. *Mater. Sci. Semicond. Process* **2023**, *165*, 2–7. [CrossRef]
17. Huang, H.; Ou, H.; Feng, J.; Du, X.; Zhang, Y. Achieving highly promoted visible-light sensitive photocatalytic activity on $BiOIO_3$ via facile iodine doping. *Colloids Surf. A Physicochem. Eng. Asp.* **2017**, *518*, 158–165. [CrossRef]
18. Yi, W.; Tang, G.; Chen, X.; Yang, B.; Liu, X. qvasp: A flexible toolkit for VASP users in materials simulations. *Comput. Phys. Commun.* **2020**, *257*, 107535. [CrossRef]
19. Debidatta, B.; Jisha, A.A.; Sharma, R.; Sanat, K.M.; Ekta, J. First Principles Study of New d^0 Half-Metallic Ferromagnetism in CsBaC Ternary Half-Heusler Alloy. *J. Supercond. Nov. Magn.* **2022**, *35*, 3431–3437.
20. Monkhorst, J.; Pack, J.D. Special Points for Brillouin-Zone Integrations. *Phys. Rev. B* **1976**, *13*, 5188–5192. [CrossRef]
21. Wu, J.; Kai, X.; Qi, Z.L.; Chen, H.Q.; Xue, M.Q.; Hui, Z.; Yu, G.; Ping, H.; Liang, J.Z. Controlling dominantly reactive (010) facets and impurity level by in-situ reduction of $BiOIO_3$ for enhancing photocatalytic activity. *Appl. Catal. B Environ.* **2018**, *232*, 136–144. [CrossRef]
22. Arora, A.; Nandi, P.; De Sarkar, A. Ferroelectricity-controlled magnetic ordering and spin photocurrent in $NiCl_2$/GeS multiferroic heterostructures. *J. Phys. Condens. Matter* **2024**, *36*, 445301. [CrossRef] [PubMed]
23. He, J.; Liu, G.; Li, X.; Wang, H.; Zhang, G. First-principles study of strain on BN-doped arsenene. *J. Mol. Model.* **2022**, *28*, 190. [CrossRef]
24. Shahriar, R.; Hoque, K.S.; Tristant, D.; Zubair, A. Vacancy induced magnetism and electronic structure modification in monolayer hexagonal boron arsenide: A first-principles study. *Appl. Surf. Sci.* **2022**, *600*, 154053. [CrossRef]
25. Qi, S.; Wu, S.; Zhang, Y.; Guan, L.; Zhang, K. Construction and first-principles analysis of BiOI and Ni doped MoS_2 Z-type heterojunctions. *J. Solid State Chem.* **2024**, *335*, 124658. [CrossRef]
26. Zhou, J.; Li, D.; Zhao, W.; Jing, B.; Ao, Z.; An, T. First-Principles Evaluation of Volatile Organic Compounds Degradation in Z-Scheme Photocatalytic Systems: MXene and Graphitic-CN Heterostructures. *ACS Appl. Mater. Interfaces* **2021**, *13*, 23843–23852. [CrossRef] [PubMed]
27. Ghaithan, H.M.; Alahmed, Z.A.; Qaid, S.M.H.; Aldwayyan, A.S. Structural, Electronic, and Optical Properties of $CsPb(Br_{1-x}Clx)_3$ Perovskite: First-Principles Study with PBE-GGA and mBJ-GGA Methods. *Materials* **2020**, *13*, 4944. [CrossRef] [PubMed]
28. Chen, Q.; Feng, X.; Teng, D.; Xu, J.; Cao, S.; Rydosz, A.; Kong, J.; Gao, F. Effect of Intrinsic Nb^{5+} Vacancy On Dielectric and Polarization Behaviors of $KSr_2Nb_5O_{15}$: First-Principles Investigation. *Phys. Status Solidi (B)-Basic Solid State Phys.* **2021**, *259*, 2100488. [CrossRef]
29. Zhao, X.; Dai, M.; Lang, F.M.; Zhao, C.; Chen, Q.Y.; Zhang, L.L.; Huang, Y.N.; Lu, H.M.; Qin, C.X. Investigating the Impact of Stress on the Optical Properties of GaN-MX2 (M=Mo, W; X=S, Se) Heterojunctions Using the First Principles. *Catalysts* **2024**, *14*, 732. [CrossRef]
30. Masihi, A.; Naseri, M.; Fatahi, N. A first-principles study of the electronic and optical properties of monolayer α-PbO. *Chem. Phys. Lett.* **2019**, *721*, 27–32. [CrossRef]
31. Ling, S.Y.; Bang, L.D.; Yu, X.H. Optical properties of arsenene nanoribbons: A first principle study. *Mater. Sci. Semicond. Process* **2021**, *136*, 106139.

coatings

MDPI

Article

Preparation and Photocatalytic Performance of Silver-Loaded Micro-Arc Oxidation TiO_2 Coating

Xingping Fan [1,*], Ying Xia [1], Wei Fan [2] and Yulong Li [3]

1 School of Vanadium and Titanium, Panzhihua University, Panzhihua 617000, China; 13183662758@163.com
2 Panzhihua University Library, Panzhihua University, Panzhihua 617000, China; fanxingping123@sina.com
3 Key Laboratory of Green Chemistry, Sichuan Institutes of Higher Education, Zigong 643000, China; yu_longli@suse.edu.cn
* Correspondence: fanxingping123@163.com

Abstract

Using a Ti6Al4V (TC4) titanium alloy plate as a substrate, micro-arc oxidation technology was applied at an oxidation time of 3.5 min and a voltage of 480 V, 495 V, and 510 V. A TiO_2–containing ceramic layer was prepared on the surface of the TC4 alloy, and the TiO_2–containing coating was doped with silver ions. The surface microstructure, phase structure, and photocatalytic performance of ceramic coatings before and after doping with silver ions were analyzed using instruments such as X-ray diffraction (XRD), energy dispersive spectroscopy (EDS), and scanning electron microscopy (SEM). The results showed that as the oxidation voltage increased, the number of large pores first decreased and then increased. At a voltage of 495 V, the total area of various pores reached 45–50 μm^2. After the voltage rose to 510 V, the maximum pore area decreased. TiO_2 exists in the surface pores of the morphology in the form of rutile, and the loading of silver ions further enhances its photocatalytic performance. The degradation rate of methyl orange by undoped silver ion samples can reach 15.5%, and the degradation rate of methyl orange can reach about 31% when 4 g/L Ag_2O is added to the electrolyte. Increasing the concentration of doped silver ions can enhance the degradation rate of methyl orange.

Keywords: micro-arc oxidation; TC4; ceramic coating; photocatalysis; TiO_2

1. Introduction

Photocatalytic technology is a technique that converts light energy into usable energy, and sunlight is also infinite. Most photocatalysts developed using photocatalytic technology are semiconductor materials, which have the advantages of a high efficiency in degrading organic matter in wastewater, simple manufacturing process, and low cost [1]. After Fujishima and Honda discovered the phenomenon of TiO_2 decomposing water under light radiation [2], many scientists began to study in this direction. In 1976, Garey et al. achieved the degradation of polychlorinated biphenyls in water using titanium dioxide [3]. Research on semiconductors in photocatalysis has gradually become a focus. Due to the high efficiency and stability of semiconductor photocatalytic conversion, scientists have begun to search for materials that can be used as photocatalysts among numerous raw materials. Among them, TiO_2 itself has become a popular material for photocatalyst research due to its excellent chemical stability, low production cost, and no toxicity [4]. Regarding research on TiO_2, previously developed powdered TiO_2 is mainly in powder form on photocatalysts. After use, it was found that the powder is difficult to recycle and can cause environmental pollution. Therefore, a reusable solid photocatalyst is sought to meet the current high requirements. Micro-arc oxidation technology can generate thin films on the surfaces of titanium alloys, magnesium alloys, and aluminum alloys, and the thin films generated from titanium alloys contain TiO_2, which has photocatalytic activity [5]. TC4, as a titanium alloy, has excellent mechanical properties and can generate TiO_2 coatings in situ on its surface through micro-arc oxidation and other methods. The coating is firmly bonded to the substrate and is not easily peeled off. The energy band of semiconductor materials consists of three parts: valence band, conduction band, and bandgap [6]. Photocatalytic reactions can convert solar radiation into chemical energy and have redox properties. When TiO_2 receives light energy greater than its bandgap energy, electrons e^- in the valence band absorb light energy and transition to the conduction band, forming electron vacancies h^+ in the original valence band, known as holes. After a series of reactions, reactive oxygen species are generated on the surface of TiO_2 [7], which can decompose metal ions and pollutants in wastewater into organic compounds. However, TiO_2 photocatalysis also has drawbacks. Firstly, the bandgap width of TiO_2 is narrow. Secondly, its photocatalytic performance is greatly influenced by the TiO_2 crystal type. Studies have found that the refractive index of anatase crystals is higher than that of rutile crystals, and the photocatalytic efficiency is higher than that of rutile crystals. However, the stability of rutile crystals is higher than that of anatase crystals, and the anatase type will irreversibly transform into the rutile type at high temperatures [8]. To further improve photocatalytic performance, there are two main approaches: one is to broaden the bandgap of TiO_2 by ion doping [9–12], and the other is to achieve stable formation of rutile TiO_2 through special means [13–15]. Although metal/TiO_2 photocatalysts have made great progress in photocatalytic reactions, there are still some shortcomings. For example, the preparation of metal/TiO_2 composite photocatalysts generally involves introducing metal components through physical or chemical means, and then immobilizing them in the form of clusters or nanoparticles on the surface of TiO_2. However, due to the lack of effective regulation of the surface interface structure between the metal and semiconductor during catalyst preparation, poor interface contact between the metal and semiconductor often occurs, seriously hindering the transfer of interface charges and weakening the promoting effect of metal components on photocatalytic reactions, thereby affecting the activity and stability of the catalyst [16,17].

There are various methods for TiO_2 modification, including metal doping, non-metal doping, dye photosensitization, semiconductor composites, and organic metal framework encapsulation [18]. These methods have a certain effect on improving the photocatalytic performance of TiO_2, but there are also some problems, such as poor thermal stability of TiO_2 photocatalysts prepared by the metal doping method, and excessive metal ions may become new electron hole pair recombination centers; the modification effect will be poor if the amount of metal ions is too low. [19]. The use of non-metallic doping to prepare photocatalysts makes it difficult for single component non-metallic doping to simultaneously expand the photoresponse range of TiO_2 materials and improve quantum efficiency. Moreover, TiO_2 doped with non-metallic elements has a weak oxidation ability, making it difficult to directly mineralize the material [20]. The sensitizers introduced by dye sensitization are mostly organic substances, which may cause secondary pollution during application. In addition, when using TiO_2 for photosensitization, TiO_2 is prone to cause photolysis of dyes [21]. The preparation of organic metal frameworks for photocatalysts requires strict laboratory environments, which limits their practical applications; moreover, its stability is poor in actual extreme environments [22]. Compared with the preparation, doping, and composite methods of other TiO_2 materials, the TiO_2 oxide film grown directly "In Situ" on the surface of the Ti alloy substrate by micro-arc oxidation (MAO) is firmly loaded, has good vibration resistance, can increase the specific surface area of the coating, and can be used for ultrasonic catalysis and photoacoustic combined catalysis of supported catalysts. It has strong processability and a great development space in the field of photocatalytic reactor design. The preparation of TiO_2 film by the MAO process can be directly designed for morphology, and is easy to modify by doping, composite, etc., making it an excellent photocatalytic material. In addition, the migration of photogenerated carriers after the deposition of precious metals will redistribute charges on the "precious metal-TiO_2", with negative charges distributed on the surface of the precious metal and positive charges distributed on the surface of TiO_2. The intermediate bandgap will create a Schottky energy barrier, forming a trap for capturing electrons and preventing the recombination of electron hole pairs. Precious metals typically deposit on the surface of TiO_2, thereby modifying semiconductors, altering charge distribution, and affecting photocatalytic performance. Shi Hongyu et al. further enhanced the photocatalytic performance of titanium dioxide particles by doping with Ag [23].

This study used TC4 as the substrate and Ag_2O as the silver ion additive to prepare a Ag–TiO_2 composite coating on the surface of the substrate using micro-arc oxidation technology. The effect of the coating on the photocatalytic degradation of methyl orange was explored, providing new ideas for the preparation of solid photocatalysts.

2. Materials and Methods

2.1. TC4 Titanium Alloy Pretreatment

The TC4 titanium alloy plate was pretreated into rectangular sheets with a size of 40 mm × 10 mm. The titanium alloy sheet was polished step by step with 600 grit, 800 grit, and 1000 grit sandpaper, and then cleaned before use.

2.2. Configuration of Electrolyte

The electrolyte is a composite system composed of sodium phosphate, sodium silicate, and sodium meta-aluminate. The formula consists of 5.5 g/L sodium silicate, 5 g/L sodium phosphate, and 0.5 g/L sodium meta-aluminate, with the addition of 4 mL/L glycerol as the base electrolyte. Then, 2 g/L, 3 g/L, and 4 g/L of Ag_2O were added to this basic electrolyte to prepare the silver-loaded micro-arc oxidation electrolyte.

2.3. Micro-Arc Oxidation Treatment

Using a TC4 titanium alloy sheet as an anode and a graphite sheet as a cathode, an AIYS750-15A pulse power (Tianjin Mingruichuang Electronic Technology Co., Ltd., Tianjin, China) supply was used for micro-arc oxidation of TC4 titanium alloy. The fixed current was 1 A; the duty cycle was 30%; the frequency was 500 Hz; the oxidation time was 3.5 min; and the oxidation voltages were 480 V, 495 V, and 510 V. A mixture of ice and water was used to cool the electrolyte during micro-arc oxidation. After oxidation, the surface was rinsed with deionized water to remove salt components from the electrolyte, and blow-dried for later use.

2.4. Photocatalytic Experiment

The experiment of photocatalytic degradation of methyl orange was conducted in a quartz test tube (70 mL). The light source used was a 300 W high-pressure mercury lamp (Shanghai Yaming Lighting Factory, Shanghai, China) with a maximum emission wavelength of 365 nm. The visible light excitation light source filtered out ultraviolet light below 380 nm through a color filter to obtain the mercury lamp. The mercury lamp was cooled by chilled water in a quartz jacket. The experiment used methyl orange as the degradation object and prepared a 10 mg/mL methyl orange solution. To prepare a 1 g/mL solution of methyl orange, weigh 0.5 g of methyl orange into a beaker; add 500 mL of deionized water to dissolve thoroughly; and transfer to a 500 mL volumetric flask for later use. To calculate the photocatalytic degradation rate of methyl orange, take 5 mL of 1 g/mL methyl orange solution in a beaker; dissolve it in 500 mL deionized water; and transfer it to a 500 mL volumetric flask. Take the 10 mg/mL methyl orange solution and measure its absorbance, denoted as A0. Take 3 culture dishes and pour 30 mL of 10 mg/mL methyl orange solution into each dish. Immerse the sample in the solution and place it in a visual colorimetric box. The distance between the lamp tube and the sample is 6–8 cm. Every half hour, use a dropper to draw a certain amount of methyl orange solution into the quartz colorimetric dish. Use a 722S visible light spectrophotometer to measure the absorbance, which is recorded as A_X. Measure the absorbance of methyl orange solution samples taken at each time point three times and take the average value. Calculate the degradation degree of methyl orange based on the relationship between absorbance and concentration.

2.5. *Material Characterization*

2.5.1. Microscopic Morphology Observation and Analysis

The surface morphology of the micro-arc oxidation ceramic film layer was characterized using a TESCAN VEGAII scanning electron microscope (TESCAN Group a.s. Brno, Czech Republic), which was fully PC-controlled and equipped with a tungsten heating filament.

2.5.2. Energy Dispersive Spectroscopy (EDS) Analysis

Using EDS energy spectrum analysis, points on the microscopic surface were selected to analyze the elemental composition, and the main roles played by the elements in combination with performance analysis were discussed.

2.5.3. XRD Phase Analysis

A DX-2700d X-ray diffractometer (Dandong Haoyuan Instrument Co., Ltd., Dandong, China) was used to detect the phase of the film layer. The measurement parameters were as follows: tube current 30 mA and voltage 40 kV, and a Cu Kα radiation in the regular range 10°~90° and a scanning speed of 8°/min. And the phase was analyzed using Powder Diffraction File (PDF4-2009, International Center for diffraction Data).

2.5.4. Distribution Pattern of Pore Area Analysis

Image J 1.8.0 was used to analyzed the pore positions from the morphology map.

2.5.5. Comparison of Photocatalytic Performance

The experiment used the degradation rate of methyl orange under photocatalysis as a control to analyze the strength of the photocatalytic performance of the sample. The degradation rate of methyl orange can be calculated using Formula (1), and was analyzed and discussed using Origin plot.

$$A = \frac{A_0 - A_X}{A_0} \times 100\% \tag{1}$$

In the formula, A represents the degradation rate of methyl orange; A_0 represents the absorbance of methyl orange before photocatalysis; and A_X represents the absorbance of methyl orange after a certain period of photocatalysis.

3. Results and Discussion

3.1. XRD Phase Analysis Results

Figure 1 shows the X-ray diffraction patterns of the samples before and after micro-arc oxidation.

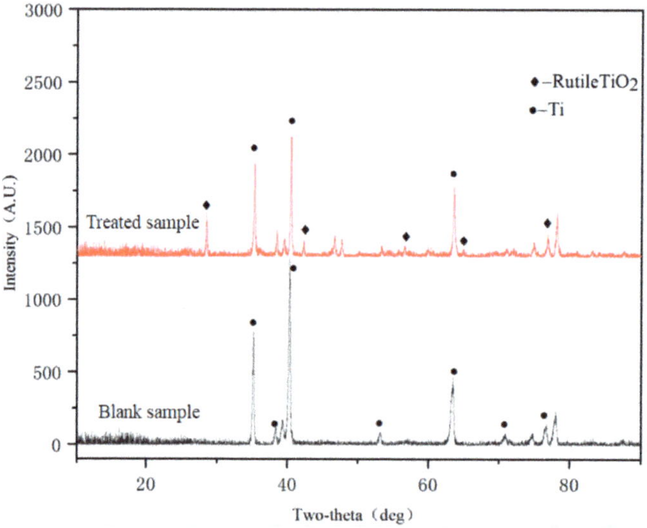

Figure 1. XRD patterns of samples before and after micro-arc oxidation.

The sample has already formed a coating during the micro-arc oxidation process. In the untreated TC4 diffraction peak spectrum, the diffraction peak at 40.23° is the diffraction peak of the titanium substrate. After micro-arc oxidation treatment, the coating is mainly composed of rutile TiO_2. The micro-arc oxidation process will undergo the following reaction [24]:

$$Ti + 2H_2O \rightarrow TiO_2 + 4H^+ + 4e^- \quad (2)$$

The reason why the spectrum did not detect rutile is because anatase TiO_2 is more easily formed at low temperatures, but rutile TiO_2 has stronger thermal stability than anatase TiO_2. At a voltage of 510 V, an electric spark is generated on the surface of the sample, causing the surrounding temperature to rise sharply to a temperature that can melt the TC4 matrix, and causing the electrolyte to vaporize. The formed anatase TiO_2 gradually transforms into rutile TiO_2 as the oxidation time increases and the temperature rises [25].

3.2. EDS Analysis

3.2.1. Composition of Undoped Silver Ion Coating Elements

Figure 2 shows the energy spectrum of the undoped silver ion coating.

After conducting element analysis on the selected points, a table was drawn as shown in Table 1. Under the condition of not doping with silver ions, the basic elements on the surface are O, Al, Si, and Ti. This indicates that O, Al, Si, and Ti are more likely to bond with TiO_2 coatings during the micro-arc oxidation process, while Na has a lower bonding force with TiO_2 coatings, and all elements exist in the form of oxides in the coating.

Table 1. Mass fraction of undoped silver ion surface elements (%).

Point	O	Al	Si	Ti	Na
6506	36.81	8.00	12.09	42.22	0.88
6508	49.54	6.18	13.88	26.84	4.06
6507	33.67	10.22	8.99	47.12	0
6509	43.55	9.54	8.44	38.47	0

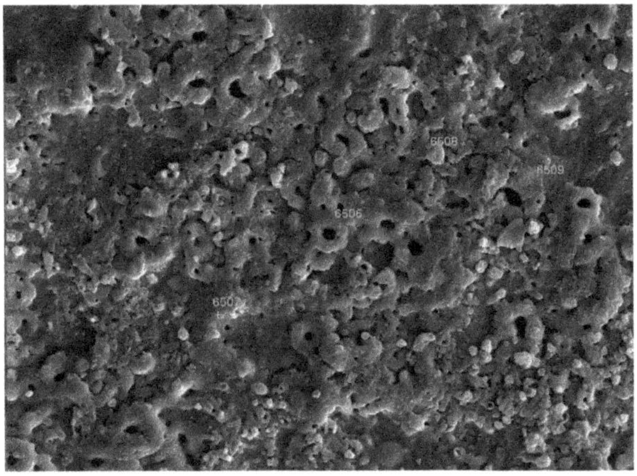

Figure 2. Point selection of undoped silver ion energy spectrum.

3.2.2. Doping Silver Ions on the Composition of Coating Elements

A point selection analysis on the sample was performed with a voltage of 510 V and an oxidation time of 3.5 min in an electrolyte containing 2 g/L Ag_2O as shown in Figure 3.

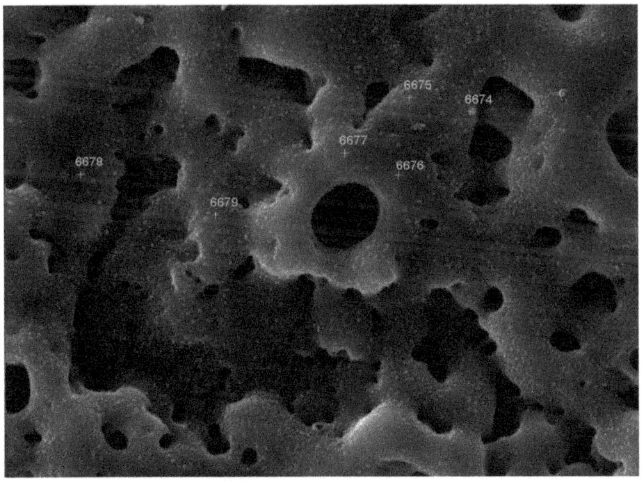

Figure 3. Energy spectrum after doping with silver ions.

The mass of surface elements doped with silver ions is shown in Table 2. It can be seen that Ag_2O, as a silver ion additive, was successfully doped and integrated into the TiO_2 coating during the micro-arc oxidation process. This EDS analysis showed an increase in the P element compared to undoped silver ion analysis. The reason is that the addition of Ag_2O increases the conductivity of the electrolyte and also enhances adhesion between the

P element and the coating. From Figure 3, it can be seen that silver ions mainly appear in the surface particles, and their forms of existence are adsorption on the coating or doping of silver ions into the TiO_2 coating, resulting in particle formation in the surface tissue.

Table 2. Mass fraction of surface elements doped with silver ions (%).

Point	O	Al	Si	Ti	P	Ag
6674	35.29	5.68	8.45	31.05	0	19.52
6675	40.66	7.03	11.52	40.79	0	0
6676	47.74	5.05	8.13	34.65	4.44	0
6677	45.15	4.68	7.88	37.24	5.05	0

3.3. Microstructure and Performance Analysis

3.3.1. Analysis of Sample Surface Morphology (SEM)

(1) Influence of voltage on the surface morphology of film layers.

Figure 4 shows the surface morphology of the samples prepared under different oxidation voltages. At different voltages, it was observed that the degree of tissue unevenness gradually became more severe as the voltage increased. The increase in voltage made the reaction of the discharge channel more intense, and the reaction time could not bring the resistance of each part of the tissue to a certain level, resulting in uneven tissue in different parts, like continuous mountain peaks, with almost no flat parts.

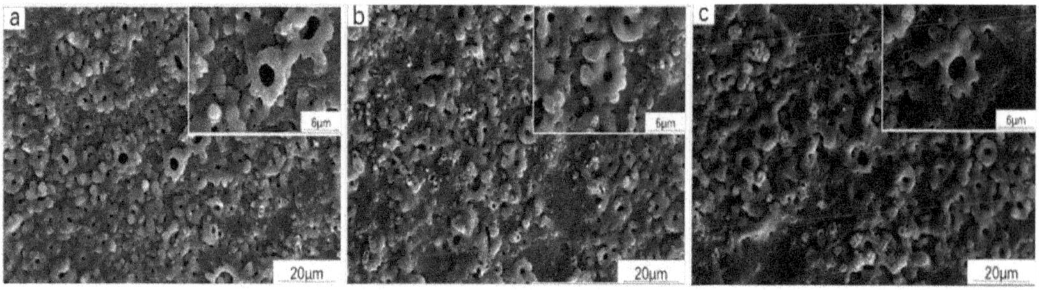

Figure 4. Surface morphology after oxidation at different voltages of (**a**) 480 V, (**b**) 495 V, and (**c**) 510 V.

3.3.2. Distribution Pattern of Pore Area

Figure 5 shows the distribution of the surface pore area of samples subjected to micro-arc oxidation at different voltages. As shown in the figure, with an increase in oxidation voltage, the number of large pores first decreases and then increases. This phenomenon is caused by the gradual increase in voltage, and the rate of breakdown to form pores is greater than the rate of material cooling to fill pores. The number of pores also increases with the increase in voltage. At a voltage of 495 V, the total area of various types of holes is as high as 45–50 μm^2. After the voltage rises to 510 V, the maximum hole area decreases, indicating that the rate of hole formation by breakdown and the rate of hole filling by material cooling also increase when the voltage rises. However, the rate of hole formation by breakdown remains higher than the rate of hole filling by material cooling.

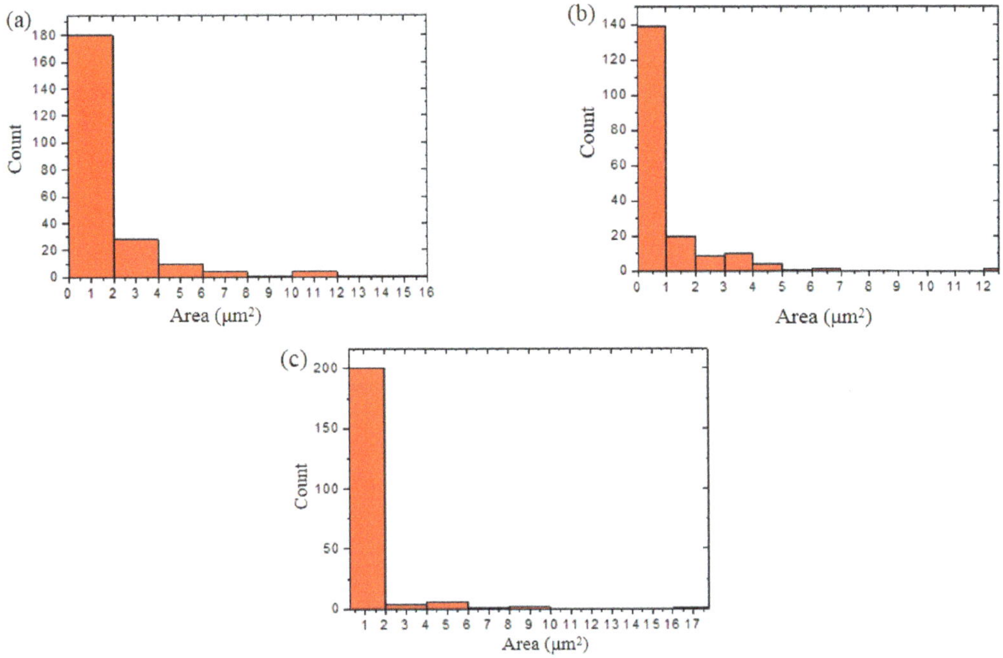

Figure 5. Statistics of hole area at (**a**) 480 V, (**b**) 495 V, and (**c**) 510 V.

3.3.3. Photocatalytic Performance

To better analyze and discuss the strength of the photocatalytic performance of the sample, the absorbance of methyl orange degradation within 3 h was measured. The degradation rate of methyl orange was calculated using a formula to compare and discuss the mechanism of photocatalytic degradation of methyl orange. Figure 6 shows the trend of the photocatalytic degradation rate of methyl orange for different samples within 3 h. With an increase in micro-arc oxidation voltage, the photocatalytic activity of the prepared film layer is enhanced. As the photocatalytic time progresses, the degradation rate of methyl orange in the same sample increases. When the time is 3 h, the highest efficiency of photocatalytic degradation of methyl orange can reach about 15.5%, and the effect is significant. This result may be related to the membrane structure on the surface of the sample.

The mechanism of photocatalytic degradation of methyl orange based on the pore area in the morphology was discussed. The pore positions from the morphology map were analyzed by Image J software. And the results are shown in Figure 7. The red represents the pore area and the green represents the non-pore area. Origin software (Origin 8.0) was used to draw a table. Table 3 shows the calculated pore area statistics. The oxidation time is 3.5 min, and the total area under different voltages is the smallest, indicating that the formation of pores is not conducive to the progress of photocatalysis. The increase in pores reduces the contact area between the surface of the sample and the methyl orange solution. TiO_2 generates reactive oxygen species ($\bullet O_2^{2-}$ and $\bullet OH$) upon light stimulation, which migrate to the surface of the sample instead of inside the pores. Therefore, an excessive pore area will reduce the performance of TiO_2 coating in producing reactive oxygen species and the area in contact with light.

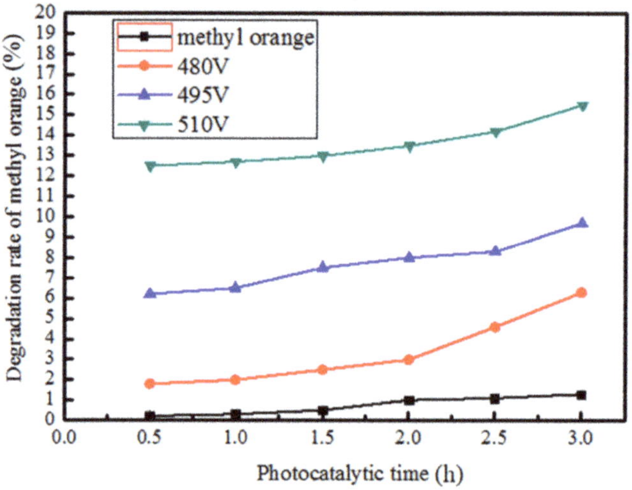

Figure 6. The effect of coatings prepared at different voltages on the degradation rate of methyl orange.

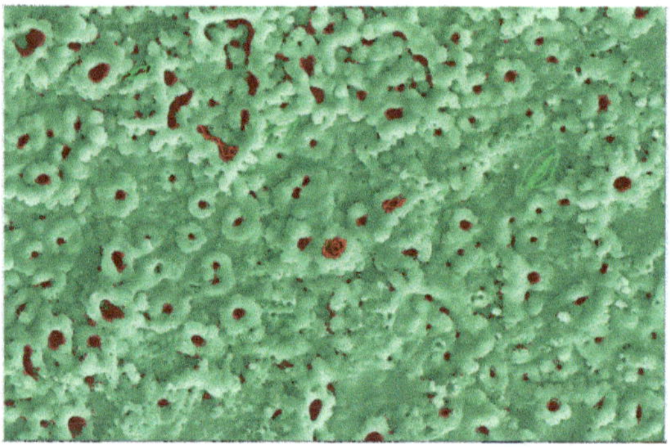

Figure 7. Image J software analysis of hole location (red area for holes, green area for non-holes).

Table 3. Statistics of hole area (μm^2).

Voltage	Total Number of Holes (N)	Mean Value
480 V	223	1.316
495 V	186	0.884
510 V	224	0.939

3.4. Influence of Doping Silver Ions on Microstructure and Properties

TiO_2 coating materials containing silver ions were prepared by adding 2 g/L, 3 g/L, and 4 g/L Ag_2O to the configured electrolyte, fixing the oxidation time for 3.5 min, and applying voltages of 480 V, 495 V, and 510 V, respectively.

3.4.1. Analysis of Surface Microstructure and Morphology of the Sample

Figure 8 shows the surface morphology of samples prepared at different voltages in a Ag_2O electrolyte containing 2 g/L. Compared with the morphology formed under different voltages, the surface tissue under 510 V voltage is smoother, with the smallest degree of concavity and convexity, but there are still many pores. Compared with the surface morphology of undoped Ag ions, there are more cohesive pores formed by the presence of small pores in the surface macropores; the most obvious of which is shown in Figure 8b at 480 V, and the smallest degree of fluctuation in the surrounding area is also shown in Figure 8b at 480 V. The reason for this phenomenon is that during the micro-arc oxidation process, continuous discharge occurs under voltage, and breakdown phenomena continue to occur, forming multiple discharge channels. The temperature of the discharge channels increases, causing the surrounding material to become molten. The molten material and electrolyte undergo a cooling reaction and solidify in the surface layer. Due to low voltage, some film layers have high resistance and cannot continue to generate electricity, causing the cooled material to push into new holes and form cohesive pores at the original hole position. Additionally, some surface layers experience film accumulation, while others remain unresponsive, resulting in varying degrees of unevenness. Under high voltage, there are no cohesive pores and the degree of fluctuation is low because the continuous breakdown of film resistance by high voltage is insufficient. In addition, after adding Ag_2O to the electrolyte, the material doped with Ag ions after cooling can quickly fill the pore positions. Therefore, as the voltage increases, the number of large pore positions in the morphology gradually decreases; the degree of fluctuation decreases; and the number of cohesive pore positions also decreases.

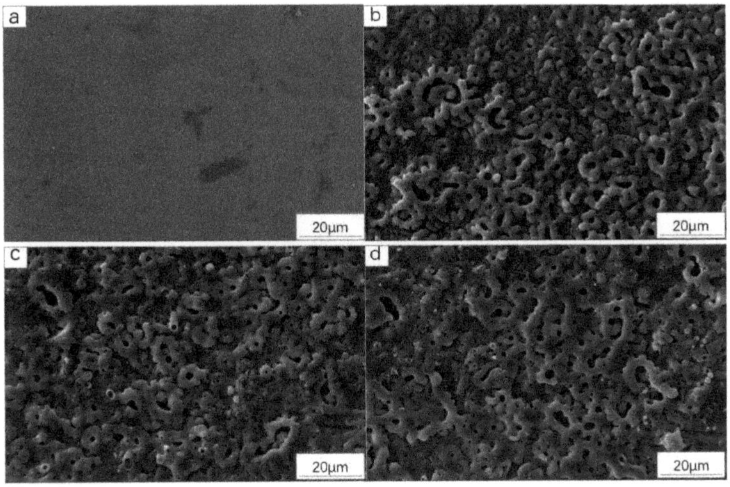

Figure 8. SEM images of the sample surface under different micro-arc oxidation voltages, 1000×: (**a**) 0 V, (**b**) 480 V, (**c**) 495 V, and (**d**) 510 V.

Figure 9 shows the morphology at 4000 times magnification. It can be seen that compared with undoped silver ion morphology, the surface tissue after doping with silver ions has more small granular tissue. Figure 9a shows a small amount of granular tissue, while Figure 9c shows the highest and most obvious amount of granular tissue, distributed in every position. Figure 9d shows that under a magnification of 10,000 times, the granular structure can be more intuitively observed. Combined with energy dispersive spectroscopy (EDS) analysis, it indicates that silver ions have been successfully doped into the TiO_2 coating and exist in a granular form on the surface. The surface tissue in Figure 9a shows the most obvious undulations, with many pores. Due to the addition of Ag_2O, under the condition of increased voltage, the undulations in Figure 9b have been improved, but the effect is not significant, and the number of pores has decreased. The undulations in Figure 9c have been significantly improved, and the number of pores has decreased significantly compared to Figure 9a, indicating that the filling rate of pores by Ag_2O increases after the voltage is increased, and the improvement in surface tissue is successful.

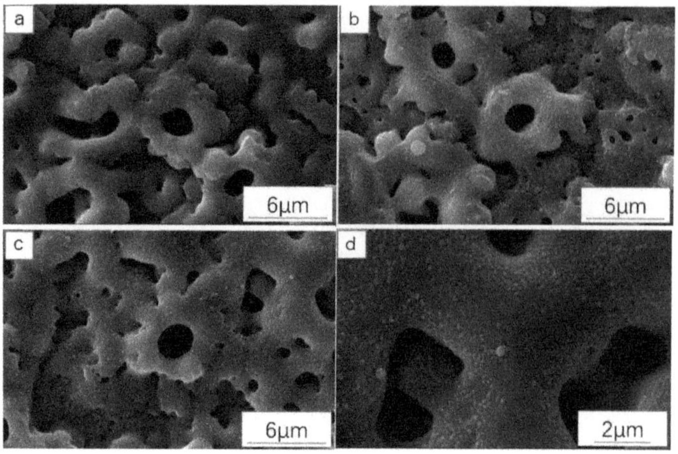

Figure 9. (**a–c**) show morphology images at 480 V, 495 V, and 510 V, respectively, at 4000× and (**d**) 10,000× times the morphology of 510 V.

3.4.2. Distribution of Pore Area After Doping with Silver

Figure 10a shows that the maximum pore area is around 30 μm^2 at 480 V. Figure 10b shows that the pore area is around 24 μm^2 at 495 V, and Figure 10c shows that the pore area returns to around 30 μm^2 at 510 V. The distribution of pore area decreases from large to small, and the number of pores decreases with an increase in voltage. The addition of Ag_2O increases the solute in the electrolyte, and the micro-arc oxidation reaction reacts violently with the increase in voltage. The solute in the electrolyte will solidify on the surface of the sample together with the molten material during the micro-arc oxidation process, and the pores will be filled. Only large holes are not filled with sufficient cooling material, resulting in irregular holes. Medium-sized holes are filled, while small holes cannot be filled with cooling material and remain.

Table 4 shows the statistical table of pore area (μm^2) after doping with 2 g/L Ag_2O at different voltages. According to the table, when the oxidation voltage is 495V, the total area of the pores is the smallest.

Table 4. Statistical table of pore area (μm^2) after doping with 2 g/L Ag_2O.

Voltage	Total Number of Holes (N)	Mean Value
480 V	350	1.380
495 V	320	1.421
510 V	270	1.793

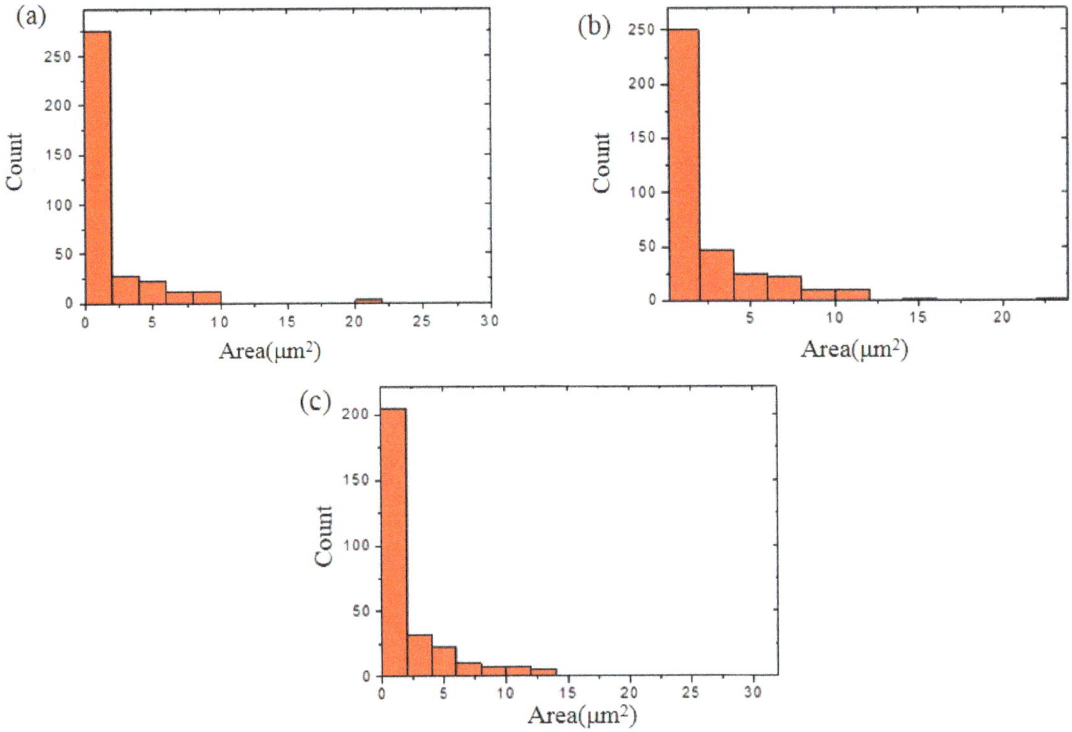

Figure 10. Distribution of pore area after doping with silver ions at (**a**) 480 V, (**b**) 495 V, and (**c**) 510 V.

3.4.3. Influence of Doping Silver Ion Concentration on Photocatalytic Performance

Figure 11 shows the degradation rate of methyl orange by samples with different doping silver ion concentrations under different voltages. Overall, for the same sample, as the concentration of doped silver ions increases, the photocatalytic effect is enhanced. When the sample obtained by micro-arc oxidation in an electrolyte containing 4 g/L Ag_2O is added, the degradation efficiency of methyl orange reaches about 31%. Compared with the blank group without doping with silver ions, the degradation rate of methyl orange was significantly improved after doping with silver ions, most notably in the sample containing 4 g/L.

Comparing the statistical table of pore area after doping with 2 g/L Ag_2O in Table 4, the total pore area of the 495 V sample is the smallest, and the degradation rate of methyl orange is also the lowest in the experimental group containing 2 g/L, indicating that the surface pore area of the sample affects the photocatalytic efficiency of the sample.

In this study, the mechanism of analysis of methyl orange degradation is as follows: Methyl orange adsorbed on TiO_2 under light irradiation absorbs visible light and is excited to inject electrons into the conduction band of TiO_2. Due to the Schottky energy barrier, excited state electrons accumulate toward silver particles through the conduction band of TiO_2 and react with adsorbed oxygen molecules on the surface of Ag to generate reactive oxygen species such as $O_2^{-}\cdot$, $HO_2\cdot$, H_2O_2, and $OH\cdot$. These reactive oxygen species then attack the methyl orange molecules in the solution and degrade them. Silver plays a role in enriching and transporting electrons on the surface of TiO_2. Due to the effect of silver atomic clusters, the reaction between the originally slow conduction band electrons of TiO_2 and oxygen molecules has become a rapid process. The process of electron transfer from the conduction band of titanium dioxide to silver atomic clusters and then the reaction with oxygen molecules is very fast, which can suppress the recombination of electrons and methyl orange cations and significantly accelerate the oxidation degradation rate of methyl orange.

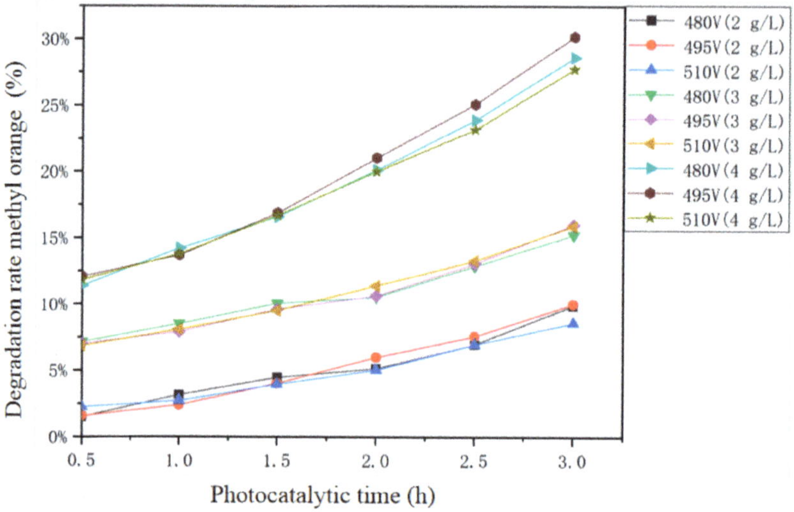

Figure 11. Degradation rate of methyl orange doped with Ag_2O.

TiO_2–coated photocatalysts loaded with silver have high stability, photocorrosion resistance, and non-toxicity. They do not produce secondary pollution during the treatment process and can be recycled and reused. From the perspective of material cycling, organic pollutants can be completely degraded into inorganic substances. Therefore, the application of TiO_2–coated photocatalysts is a clean treatment technology.

4. Conclusions

A coating mainly composed of rutile TiO_2 was formed on the surface of TC4 by micro-arc oxidation treatment. Voltage has the greatest impact on the film layer, and an increase in voltage will continuously break down the film layer, forming volcanic pores and continuous structures. The total area of pores formed with an oxidation time of 3.5 min is the smallest, and after doping with Ag_2O, the total area of pores increases. Before doping with Ag_2O, the highest photocatalytic degradation rate of methyl orange was only 15.5%. Increasing the concentration of doped silver ions improves the degradation rate of methyl orange. After doping with Ag_2O, the highest photocatalytic degradation rate of methyl orange could reach about 31%, indicating that silver loading on TiO_2 coatings can improve their photocatalytic performance.

Author Contributions: Experiments, X.F.; writing, X.F. and Y.X.; editing and translation, W.F. and Y.L. All authors have read and agreed to the published version of the manuscript.

Funding: This research was funded by the Opening Project of Key Laboratory of Green Chemistry of Sichuan Institutes of Higher Education No: LYJ2302.

Institutional Review Board Statement: Not applicable.

Informed Consent Statement: Not applicable.

Data Availability Statement: Data are contained within the article.

Conflicts of Interest: The authors declare no conflicts of interest.

References

1. Xue, S.S. Research Progress in titanium dioxide doped modification and photocatalytic degradation of pollutants. *Mod. Chem. Res.* **2023**, *16*, 19–21.
2. Fujishima, A.; Honda, K. Electrochemical photolysis of water at a semiconductor electrode. *Nature* **1972**, *238*, 37–38. [CrossRef] [PubMed]
3. Carey, J.H.; Lawrence, J.; Tosine, H.M. Photodechlorination of PCB's in the presence of titanium dioxide in aqueous suspensions. *Bull. Environ. Contam. Toxicol.* **1976**, *16*, 697–701. [CrossRef]
4. Arun, J.; Nachiappan, S.; Rangarajan, G. Synthesis and application of titanium dioxide photocatalysis for energy, decontamination and viral disinfection: A review. *Environ. Chem. Lett.* **2023**, *21*, 339–362. [CrossRef] [PubMed]
5. Li, P.L. Preparation and Properties of MAO Coating on Titanium Fiber Surface. Master's Thesis, Chang'an University, Xi'an, China, 2023.
6. Zhang, B.W. Preparation of Modified TiO_2-Based Photocatalyst and Its Photocatalytic Degradation of Organic Pollutants. Master's Thesis, Guangxi University, Nanning, China, 2023.
7. Wan, R.S. Photocatalytic Degradation of Formaldehyde and Deactivation-Regeneration of Supported TiO_2. Master's Thesis, Nanchang University, Nanchang, China, 2023.
8. Li, Y.Z. Preparation and Photocatalytic Performance of Tungsten Doped TiO_2 Nanotubes on the Surface of Titanium Mesh. Master's Thesis, Liaoning Normal University, Dalian, China, 2023.
9. Hung, R.Y.; Luo, X.Y.; Zhao, D.F. Preparation and photocatalytic properties of Ag doped TiO_2. *Nonferr. Met. Met. Sci. Eng.* **2016**, *7*, 67–72.
10. Yu, L.Z.; Li, J.W.; Lin, Y.H. Research progress on the photocatalytic property of modifiednano-TiO_2 doped with transition metal ions. *Energy Chem. Ind.* **2019**, *40*, 11–17.
11. Li, B.; Wang, X.M.; Wei, H.M. Effect of B-F gradient doping on photocatalytic properties of TiO_2 thin films. *J. Qinghai Univ.* **2018**, *36*, 93–98.
12. Lee, J.H.; Lee, Y.K.; Kim, Y.J. Surface characteristics and photocatalytic propertiy of B doped TiO_2 layer synthesized by plasma electrolytic oxidation process. *Korean J. Mater. Res.* **2021**, *31*, 552–561. [CrossRef]
13. Gao, X.; Feng, Q.; Fu, Y. Mechanism of microscopic adsorption of organic molecules with different functional groups on the surface of anatase phase TiO_2(101). *J. At. Mol. Phys.* **2023**, *40*, 31–38.
14. Yu, H.; Yao, Z.; Yu, Z.; Tian, F.Z.; Su, Y.X.; Wang, Y.B.; Yu, J. Application of black titanium dioxide and its composite materials in photocatalytic degradation of water pollutants. *Fine Chem.* **2025**, *1*, 1–18.
15. Diana, R.; Geometry, A.; Muhamad, D. Study on triphase of polymorphs TiO_2 (Anatase/rutile/brookite) for boosting photocatalytic activity of metformin degradation. *Chemosphere* **2024**, *351*, 141206.
16. Wang, H.; Wang, L.; Lin, D. Titanosilicate zeolite supported Pt nanoparticles with electronic metal-supportinteractions for efficient methanol steam reforming. *Catal. Today* **2021**, *382*, 42–47. [CrossRef]
17. Wu, X.K.; Wang, Z.C.; Zhang, D. Solvent-free microwave synthesis of ultra-small Ru-Mo_2C@CNT with strong metal-support interaction for industrial hydrogen evolution. *Nat. Commun.* **2021**, *12*, 4018–4027. [CrossRef]
18. Wang, B.; Sa, G.L.; Xu, A.J. Progress of photocatalytic degradation of organic pollutants over TiO_2. *Ind. Catal.* **2020**, *28*, 20–24.
19. Guo, J.R.; Han, F.; Li, C.R.; Deng, H.Z.; Li, J. Study on the photocatalytic degradation of organic dye by Cu-doped TiO_2. *Appl. Chem. Ind.* **2016**, *45*, 653–656.
20. Zhou, J.W.; Wei, Z.B.; Sun, M.; Hao, Y.Y.; Huang, J.; Zhang, Y.; Liu, R.X. Enhancement of Photocatalytic Activity via Boron-doped Titanium Dioxide. *J. Xinxiang Univ.* **2019**, *36*, 11–16.
21. Wen, Y.X.; Bao, L.H.; Ling, J.W. Photocatalytic activity of dye-sensitized titanium dioxide/graphene composite. *Dye Finish* **2021**, *47*, 7–12.
22. Zhang, Q.Y.; Wu, L.M.; Tian, Y.Y.; Yu, R.F.; Zhang, Y.T. Preparation and photocatalytic properties of MOF-derived TiO_2/heteropolyacids pomposites. *Technol. Water Treat.* **2022**, *48*, 71–75.
23. Shi, H.Y.; Li, G.Z.; Li, J.M.; Zhang, S.L.; Luo, L.J.; Wang, H.B.; Yang, M. Study on photocatalytic degradation of methylene blue by silver doped titanium dioxide. *J. Yunnan Minzu Univ. (Nat. Sci. Ed.)* **2024**, *33*, 314–319.
24. Xu, Y.K.; Luo, Y.Q.; Jiang, J.L. Influence of electrolyte Ca/P ratio on microstructure and properties of micro-arc oxidation titanium alloy coatings. *Rare Metal. Mat. Eng.* **2023**, *52*, 675–684.
25. Li, L.X.; Zhang, R.Z.; Liu, Z.P. Catalytic properties of different crystalline TiO_2 for ozoneoxidation. *New Chem. Mat.* **2024**, *52*, 184–189.

coatings

MDPI

Article

Production of Self-Supporting Hollow Carbon Nanofiber Membranes with Co/Co_2P Heterojunctions via Continuous Coaxial Co-Spinning for Efficient Overall Water Splitting

Ruidan Duan [1,†], Jianhang Ding [1,†], Jiawei Fan [1] and Linzhou Zhuang [1,2,*]

1 State Key Laboratory of Chemical Engineering, School of Chemical Engineering, East China University of Science and Technology, Shanghai 200237, China; y30230195@mail.ecust.edu.cn (R.D.); y30240045@mail.ecust.edu.cn (J.D.); y30240047@mail.ecust.edu.cn (J.F.)

2 Shanghai Key Laboratory of Multiphase Materials Chemical Engineering, East China University of Science and Technology, Shanghai 200237, China

* Correspondence: lzzhuang@ecust.edu.cn

† These authors contributed equally to this work.

Abstract

To address mass transport limitations in carbon nanofiber membrane electrodes for overall water splitting, a self-supporting nitrogen-doped hollow carbon nanofiber membrane embedded with Co/Co_2P heterojunctions (Co/Co_2P-NCNFs-H) was fabricated via continuous coaxial electrospinning. The architecture features uniform hollow channels (200–250 nm diameter, 30–50 nm wall thickness) and a high specific surface area (254 m^2 g^{-1}), as confirmed by SEM, TEM, and BET analysis. The Co/Co_2P heterojunction was uniformly dispersed on nitrogen-doped hollow carbon nanofibers through electrospinning, leverages interfacial electronic synergy to accelerate charge transfer and optimize the hydrogen evolution reaction (HER) and the oxygen evolution reaction (OER). Electrochemical tests demonstrated exceptional catalytic activity, achieving current densities of 100 mA cm^{-2} at ultralow overpotentials of 405.6 mV (OER) and 247.9 mV (HER) in 1.0 M KOH—surpassing most reported transition metal catalysts for both half-reactions. Moreover, the electrode exhibited robust long-term stability, maintaining performance for nearly 20 h at 0.6 V (vs. Ag/AgCl) (OER) and over 250 h at −1.5 V (vs. Ag/AgCl) (HER), attributed to the mechanical integrity of the hollow architecture and strong metal–carbon interactions. This work demonstrates that integrating hollow nanostructures (enhanced mass transport) and heterojunction engineering (optimized electronic configurations) creates a scalable strategy for designing efficient bifunctional catalysts, offering significant promise for sustainable hydrogen production via water electrolysis.

Keywords: water electrolysis; self-supporting hollow carbon nanofibers; continuous coaxial co-spinning; heterojunction engineering; transition metal phosphides

1. Introduction

Limited reserves of fossil fuels and their severe environmental impacts have prompted an urgent search for clean and sustainable renewable energy alternatives [1,2]. Hydrogen energy has emerged as one of the cleanest energy carriers, attracting increasing attention due to its environmental friendliness, high efficiency, and zero-emission characteristics [3–5]. Among various hydrogen production technologies, water electrolysis stands out as the cleanest approach [5,6], offering high-purity hydrogen with zero carbon emissions and excellent compatibility with renewable energy sources such as solar and wind power. However, the two half-reactions involved in overall water electrolysis—-the hydrogen evolution reaction (HER) and the oxygen evolution reaction (OER)—suffer from sluggish kinetics, necessitating highly efficient catalysts to enhance the reaction rates and overall efficiency of water electrolysis systems [7,8]. This sluggishness translates to significant overpotentials—the extra voltage beyond the thermodynamic requirement (1.23 V vs. RHE for overall water splitting)—needed to drive the reactions at practical current densities. High overpotentials directly reduce the overall energy efficiency of the system [9]. For instance, many current systems require overpotentials exceeding 300 mV for the OER and 100 mV for the HER to achieve even a modest current density of 10 mA cm^{-2}, a common benchmark for initial activity assessment [8,10].

In recent years, significant efforts have been devoted to developing highly active catalysts for HER and OER. Noble-metal-based electrocatalysts exhibit excellent activity and durability [11]; for example, Pt-based materials remain the most effective HER catalysts, often achieving current densities of 10 mA cm^{-2} at overpotentials below 50 mV [12], and Ir/Ru oxides exhibit optimal OER performance, typically requiring overpotentials of 250–350 mV to deliver 10 mA cm^{-2} [13,14]. While these materials demonstrate high efficiency and good stability, their large-scale application is severely constrained by their high cost and limited natural abundance [15]. This has necessitated the development of non-noble metal catalysts, particularly cost-effective, durable bifunctional catalysts. The goal is to identify materials that not only drive both the HER and the OER efficiently, but also achieve this at industrially relevant current densities (e.g., >100 mA cm^{-2} or even >500 mA cm^{-2}) while minimizing overpotential. Furthermore, long-term operational stability, often defined as maintaining performance for hundreds or thousands of hours without significant degradation, is a critical metric for practical viability. Such catalysts could significantly simplify device fabrication and operation while reducing hydrogen production costs, offering substantial commercial potential [16,17].

Recent advances have yielded numerous non-noble metal and even metal-free electrocatalysts, including transition metal (Fe, Co, Ni, Mn, Mo, etc.) oxides/hydroxides [18,19], carbides [20], nitrides [21], phosphides [22], and sulfides [23]. Among these, transition metal phosphides (e.g., NiP [24], FeP [25], MoP [26], and CoP [27]), especially CoP and Co_2P, demonstrate exceptional HER/OER activity and good electrical conductivity [28]. In addition, rational design of heterointerfaces has emerged as a powerful strategy for electrocatalyst optimization. The strong electronic coupling interactions between different phases in these heterogeneous structures can effectively modulate electronic configurations, reduce reaction energy barriers, and optimize interactions with reaction intermediates, thereby promoting chemisorption of reactive species on the catalyst surface and significantly boosting electrocatalytic activity [29–31]. Zhong et al. [31] exemplified this approach with Co/CoP@HOMC, where interfacial charge redistribution and hierarchical porosity synergistically achieve ultralow overpotentials (η_{10} = 120 mV for HER/260 mV for OER) in alkaline media. While promising, these materials often face challenges in simultaneously achieving low overpotentials for both reactions, high current densities, and prolonged stability. For example, many phosphides might show good HER activity but require higher OER overpotentials, or vice versa, and stability can be compromised under harsh oxidative conditions for OER or reductive conditions for HER over extended periods [32,33].

Furthermore, the catalytic performance of transition metal phosphides can be further enhanced through synergistic interactions with self-supporting graphitic carbon materials, which improve both electronic conductivity and structural stability [34,35]. Carbon nanofibers have emerged as particularly promising supports due to their rapid electron transport and excellent electrochemical stability [36]. Heteroatom doping, particularly nitrogen incorporation, plays a crucial role in modifying the electronic structure of carbon supports, creating intrinsic active sites, and improving overall electrocatalytic performance [37]. Meanwhile, nitrogen doping enhances the three-dimensional hydrophilicity of catalyst surfaces, facilitating electrolyte–electrode contact and boosting catalytic efficiency. Wei et al. [38] successfully constructed a heterostructure interface between Fe/Ni phosphides and graphitic carbon nitride (C_3N_4). The nitrogen species in C_3N_4 effectively modulates the electronic structure of Fe/Ni sites and optimizes their adsorption strength with oxygen-containing intermediates. The resulting FeNi-C_3N_4-P catalyst demonstrates exceptional OER performance, achieving a current density of 100 mA cm^{-2} at an ultralow overpotential of merely 235 mV. Beyond composition, nanostructural design critically determines catalytic performance [39]. Hollow porous carbon nanofibers, with their well-defined internal channels, high surface area, optimal aspect ratio, and superior conductivity, provide abundant active sites while minimizing electron/mass transport distances, accelerating electrolyte diffusion and gas bubble release [40,41]. These features are critical for achieving high current densities by ensuring efficient reactant supply and product removal and can contribute to improved stability by preventing electrode fouling or active site blockage [42]. Core-shell nanostructures with excellent stability, functionality, and dispersion have proven effective for enhancing electrocatalysis [43,44]. Electrospun core-shell nanofibers offer particularly precise control over active site regulation and charge transfer resistance reduction. Li et al. [45] developed a hybrid CoP catalyst anchored on nitrogen-doped hollow carbon spheres supported by carbon nanofibers (CoP/NCF-200) through controlled electrospinning followed by pyrolysis and phosphidation. This catalyst exhibited remarkable activity, favorable kinetics, and excellent stability, attributable to both the synergistic effects between CoP and N-doped carbon and the unique hollow spherical architecture. Therefore, transition metal co-doped carbon catalysts are widely regarded as promising alternatives to noble-metal-based catalysts for water electrolysis.

Several methods, including hydrothermal growth [46], templating with Metal-Organic Frameworks (MOFs) [47], and vacuum filtration [48], have been developed to fabricate self-supporting electrocatalysts. However, electrospinning stands out as a particularly versatile and scalable technique for creating continuous one-dimensional (1D) nanofibrous membranes. A key advantage of this method is the direct production of binder-free, interwoven 3D networks that possess both high mechanical strength and excellent electrical conductivity pathways throughout the material [49]. Furthermore, the use of coaxial electrospinning provides precise control over the fabrication of complex core-shell nanostructures. This capability is essential for rationally designing advanced architectures, such as the hollow fibers presented in this work, where a sacrificial core material can be selectively removed during post-processing to create internal channels that enhance mass transport [50]. While the process requires careful optimization of multiple parameters to achieve a desired morphology, its ability to uniformly embed catalyst precursors within the polymer matrix ensures homogeneous distribution of active sites, making it a superior strategy for synthesizing high-performance, self-supporting electrodes compared to methods that may suffer from particle aggregation or non-uniform growth on a substrate [49].

To address the aforementioned challenges of achieving low overpotentials, high current densities, and robust long-term stability in bifunctional catalysts, particularly concerning mass transport limitations and active site optimization, here, a self-supporting nitrogen-doped hollow carbon nanofiber membrane embedded with Co/Co_2P heterojunctions (Co/Co_2P-NCNFs-H) was fabricated via continuous coaxial co-spinning. The inner polymethyl methacrylate (PMMA) layer was pyrolyzed during high-temperature carbonization, creating hollow channels that promote gas diffusion and expose abundant active sites. The outer layer, composed of cobalt precursors and polyacrylonitrile (PAN), formed a conductive carbon matrix after carbonization. The Co/Co_2P heterojunction leverages interfacial electronic synergy to accelerate charge transfer and enhance bifunctional catalytic activity. Electrochemical tests in 1.0 M KOH revealed exceptional performance, achieving a high current density of 100 mA cm^{-2} at competitive overpotentials of 405.6 mV (OER) and 247.9 mV (HER) in 1.0 M KOH, alongside robust stability with continuous operation for over 250 h. These metrics address key challenges in the field. This work highlights the critical role of hollow nanostructures in optimizing mass transport and heterojunction engineering in designing high-efficiency electrocatalysts, offering a scalable strategy for advancing sustainable hydrogen production technologies.

2. Experimental Section

2.1. *Materials*

Polyacrylonitrile (PAN, Mw = 150,000 g mol^{-1}, AR grade) and polymethyl methacrylate (PMMA, AR grade) were purchased from Aladdin Reagent Inc., Shanghai, China. Cobalt(II) acetate tetrahydrate ($Co(AC)_2 \cdot 4H_2O$, 99.9%), phosphoric acid (H_3PO_4, 99.9%), zinc acetate ($Zn(AC)_2 \cdot 2H_2O$, 99.9%) and N, N-dimethylformamide (DMF, AR grade) were purchased from Shanghai Titan Scientific Co., Ltd., Shanghai, China.

2.2. *Preparation of Self-Supporting Co/Co_2P-NCNFs-H Membranes*

Synthesis of Co/Co_2P-NCNFs-H was performed via coaxial electrospinning followed by a multi-step heat treatment, as illustrated in Figure S1. The procedure is detailed below.

2.2.1. Preparation of Precursor Solutions

Outer Shell Solution (Solution A): First, 0.1 g of $Co(AC)_2 \cdot 4H_2O$ (0.4 mmol) was dissolved in 15 mL of N,N-dimethylformamide (DMF) and sonicated to form a homogeneous solution. Subsequently, 0.2 g of H_3PO_4 was added, followed by 1.5 g of polyacrylonitrile (PAN). The mixture was stirred overnight to yield a uniform, pink spinning solution.

Inner Core Solution (Solution B): In a separate vessel, 0.5 g of $Zn(AC)_2 \cdot 2H_2O$ and 1.0 g of polymethyl methacrylate (PMMA) were dissolved in 15 mL of DMF by stirring in a 50 °C water bath. After a homogeneous solution was formed, 1.0 g of PAN was added, and the mixture was stirred overnight at room temperature to obtain the inner core solution.

2.2.2. Coaxial Electrospinning

The two precursor solutions were placed into separate 10 mL syringes and loaded onto two syringe pumps. Electrospinning was performed under the following conditions:

Setup: A coaxial spinneret consisting of a 17-gauge outer needle (for Solution A) and a 21-gauge inner needle (for Solution B).

Flow Rate (Outer Solution A): 1.0 mL h^{-1}.

Flow Rate (Inner Solution B): 0.5 mL h^{-1}.

Applied Voltage: 19 kV.

Collector Distance: 15 cm. The process was run for 6 h to produce a uniform nanofiber membrane, denoted as Co/H_3PO_4-PAN.

2.2.3. Post-Treatment: Pre-Oxidation, Carbonization, and Activation

Pre-oxidation: The as-spun Co/H_3PO_4-PAN membrane was first dried in a vacuum oven at 60 °C to remove residual DMF. It was then transferred to a muffle furnace and heated to 280 °C at a ramp rate of 1 °C min^{-1} and held for 1 h to stabilize the fiber structure. This yielded the pre-oxidized sample, Co/H_3PO_4-PAN-O.

Carbonization: The stabilized membrane was placed in a tube furnace and carbonized under an argon atmosphere. The furnace was heated to 900 °C at a ramp rate of 5 °C min^{-1}

and held for 3 h. During this step, the inner PMMA and zinc components were pyrolyzed, resulting in the formation of porous hollow carbon nanofibers.

Activation: The final carbonized sample was immersed in 0.5 M H_2SO_4 for 2 h for activation, then washed thoroughly with deionized water and dried. The final product was labeled Co/Co_2P-NCNFs-H.

2.3. Preparation of Control Samples

Co/Co_2P-NCNFs (Solid Fibers): This control sample was fabricated using single channel electrospinning. Only the outer shell solution (Solution A) was used, with an injection rate of 1.0 mL h^{-1}. All subsequent pre-oxidation and carbonization steps were identical to those described in Section 2.2.3.

Co-NCNFs-H (Hollow, P-free Fibers): This control was prepared to investigate the effect of phosphorus. Fabrication followed the coaxial electrospinning procedure described in Section 2.2.2, using the same inner core solution (Solution B). However, the outer shell solution was prepared without the addition of H_3PO_4. All other processing parameters and post-treatments remained constant.

Co-NCNFs (Solid, P-free Fibers): This solid control sample was prepared by single-channel electrospinning using the P-free outer solution described above at an injection rate of 1.0 mL h^{-1}. The post-treatment conditions were identical to those for the other samples.

2.4. Catalytic Activity Measurements

Electrochemical measurements were conducted using a CHI 660E electrochemical workstation (CHI Instruments, Shanghai, China) in a three-electrode setup with a 1.0 M KOH electrolyte solution at room temperature and ambient pressure. As-prepared Co/Co_2P-NCNFs-H was employed as a self-supported working electrode with an exposed area of 1 cm^2. An Ag/AgCl electrode and a glassy carbon electrode served as the reference and counter electrodes, respectively. Cyclic voltammetry (CV) was performed in the 0~1.1 V range at a scan rate of 5 mV s^{-1} to evaluate water oxidation (OER). The linear sweep voltammetry (LSV) method was applied in the −1~−2 V range to obtain the hydrogen evolution reaction (HER) performance with a scan rate of 5 mV s^{-1}. The stability of the catalysts was assessed using the chronoamperometry technique. All potential values were converted to the reversible hydrogen electrode according to Equation (1), and the overpotential (η) was calculated according to Equation (2):

$$E\ (RHE) = E\ (Ag/AgCl) + 0.059 \times pH + 0.197 \tag{1}$$

$$\eta = E(RHE) - 1.23 \tag{2}$$

All electrochemical data were i-R corrected (0.8 iR).

ECSA evaluation was conducted using the CV method in the non-faradaic current region at scan rates of 20, 40, 60, 80, and 100 mV s^{-1}. The current density difference at selected potentials was plotted versus scan rate. The resulting linear slope equaled C_{dl}. ECSA was calculated according to Equation (3), using the specific capacitance (C_s) of a standard electrode with a geometric surface area of 1 cm^2. The Cs typically ranges from 20 to 60 μF cm^{-2}. Herein, an average value of 40 μF cm^{-2} was used to calculate the ECSA.

$$ECSA = C_{dl}/C_s \tag{3}$$

Turnover frequency (TOF) was calculated according to Equation (4), where I is the current density at a given overpotential (e.g., η = 200 mV), F is the Faraday constant (96,485 C), and n is the molar quantity of electrochemically active sites.

$$TOF = I/(4F \times n) \quad (4)$$

Electrochemical impedance spectroscopy (EIS) measurements were performed in the frequency range of 100 kHz to 0.01 Hz, and OER/HER was tested at 0.6 V/−1.2 V vs. Ag/AgCl, respectively. The relevant R_{ct} values were fitted using ZView2 software. A smaller R_{ct} value suggested that the catalyst has a higher charge transfer rate.

3. Results and Discussion

3.1. Morphology Characterization

The morphology and microstructure of the synthesized samples were systematically characterized. Scanning electron microscopy (SEM) was employed to analyze the fibrous structure of the as-prepared Co/Co_2P-NCNFs-H, as illustrated in Figure 1a. The SEM images revealed a tightly interwoven three-dimensional network of nanofibers, ensuring continuous conductive pathways. The fibers exhibited uniform diameters ranging from 200 to 250 nm, with submicron-scale dimensions contributing to their exceptional mechanical strength [51]. Cross-sectional SEM images (Figure 1b,c) confirmed the hollow architecture, attributed to the thermal decomposition of the polymethyl methacrylate (PMMA) core and the phase separation between PMMA and polyacrylonitrile (PAN). The removal of PMMA during carbonization exposed abundant active sites, thereby enhancing catalytic performance. The dual presence of PAN in both inner and outer layers facilitated robust integration of active sites with the carbon shell, effectively preventing metal particle detachment and ensuring structural stability. Notably, the fibers displayed no observable agglomeration, with wall thicknesses uniformly maintained at 30–50 nm. The self-supporting, three-dimensional network of nanofibers provides continuous pathways for electron transport and ensures robust mechanical integrity. Critically, the hollow interior of each nanofiber is intentionally designed to create channels that facilitate rapid electrolyte diffusion to the active sites and efficient release of generated gas bubbles (O_2 and H_2). This structure increases the overall specific surface area, thereby maximizing the exposure of catalytically active sites [41].

The fine microstructure of the Co/Co_2P-NCNFs-H sample was characterized by transmission electron microscopy (TEM). The TEM images clearly reveal the hollow structure of the carbon nanofibers, with no observable metal agglomeration (Figure 1d,e), demonstrating that the N,P-doped carbon matrix effectively promotes uniform dispersion of Co species. High-resolution TEM (HRTEM) analysis showed lattice fringes of 0.206 nm, which belong to the (211) plane of Co_2P (Figure S2) [52], but no obvious Co and C lattice is observed, indicating the predominantly amorphous nature of the C and Co components (Figure 1f). This may be due to the catalyst particles being too small and evenly distributed. Furthermore, energy-dispersive X-ray spectroscopy (EDX) elemental mapping confirmed the homogeneous distribution of multiple elements (C, N, P, O, and Co) throughout the Co/Co_2P-NCNFs-H sample (Figure 1g). The close proximity of Co and P atoms is expected to promote formation of Co/Co_2P heterojunctions, which are anticipated to lower reaction energy barriers and accelerate charge transfer due to synergistic interfacial electronic effects [28]. Furthermore, uniform doping of nitrogen into the carbon framework is expected to modulate the electronic structure, improve surface hydrophilicity for better electrolyte contact, and securely anchor active species to enhance long-term stability.

SEM cross-sectional and TEM images of the control sample Co/Co_2P-NCNFs are shown in Figure S3. Unlike the Co/Co_2P-NCNFs-H sample, the SEM cross-sectional image of Co/Co_2P-NCNFs can be observed as a solid carbon nanofiber structure. From the TEM images, it can be observed that there are carbon shell-encapsulated nanoparticles loaded onto the carbon nanofibers, and the diameter of the particles is between 10~15 nm

(Figure S3b,c). The EDS patterns show that the elements C, N, P, O, and Co are distributed in the nanofibers in the Co/Co_2P-NCNFs, but the elements Co and P are more concentrated in the nanoparticles, and the nanoparticles are mainly Co/Co_2P heterojunctions coated by the outer carbon shell (Figure S4). Through characterization of the microscopic morphology of Co/Co_2P-NCNFs-H and control samples, it can be proved that reasonable design of coaxial co-spinning conditions can achieve preparation of hollow carbon nanofibers doped with N and P elements, and that a reasonable flow rate of the inner and outer layers of the solution can lead to more homogeneous dispersion of the Co element during the electrospinning process.

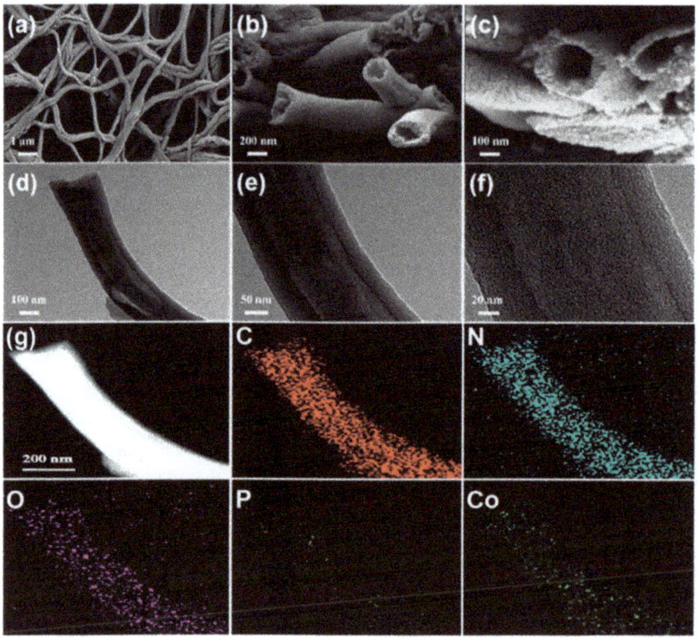

Figure 1. Morphology characterization of Co/Co_2P-NCNFs-H. (**a–c**) SEM images of Co/Co_2P-NCNFs-H. (**d,e**) TEM images of Co/Co_2P-NCNFs-H. (**f**) HRTEM images of Co/Co_2P-NCNFs-H. (**g**) HAADF-STEM image and EDS elemental mapping of Co/Co_2P-NCNFs-H.

3.2. Composition Study of the Samples

XRD was used to characterize the degree of graphitization and crystal structure of the catalysts. A broad peak located at 26°, representing the amorphous structure of carbon, was detected in all four samples (Figure 2a). Three distinct sharp peaks were detected in both the Co-NCNFs-H and Co-NCNFs samples at 44.2°, 51.5°, and 75.8°, which corresponded to the Co(111), Co(200), and Co(220) crystallographic planes, respectively (JCPDS No. 15-0806) [53]. There are two distinct Co_2P peaks in Co/Co_2P-NCNFs-H and Co/Co_2P-NCNFs samples located at 43.29° and 40.72°, corresponding to the Co_2P (211) and Co_2P (121) crystal faces (JCPDS No. 32-0306) [54]. Also, weak peaks of Co (111), Co (200), and Co (220) were present in the XRD pattern of the sample, representing the coexistence of metal Co and Co_2P within the carbon nanofiber matrix. The small crystallite sizes and lattice strain likely contribute to enhanced catalytic activity via increased density of grain boundaries and interfacial defects, which can serve as additional active sites [55].

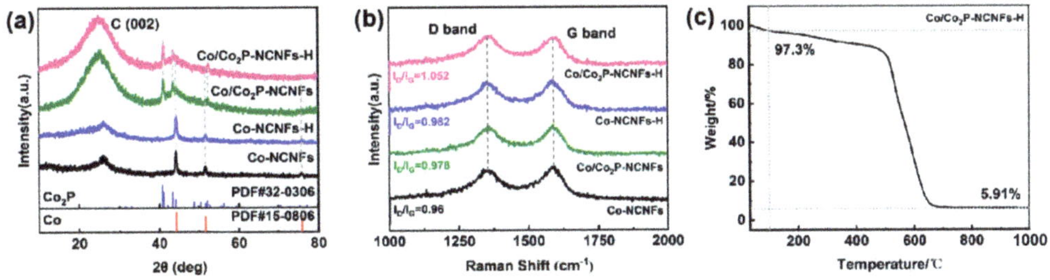

Figure 2. Composition study of the samples. (**a**) XRD patterns and (**b**) Raman spectra of Co-NCNFs, Co/Co_2P-NCNFs, Co-NCNFs-H, and Co/Co_2P-NCNFs-H. (**c**) TGA analysis of Co/Co_2P-NCNFs-H.

Meanwhile, Raman spectroscopy was employed to probe the structural characteristics and defect states of the carbon nanofiber framework (Figure 2b). All samples exhibit two prominent peaks: the D band at ~1350 cm^{-1}, attributed to disordered carbon or defects, and the G band at ~1587 cm^{-1}, corresponding to graphitized carbon atoms [56]. The ratio of the D band to the G band (I_D/I_G) is used to determine the degree of graphitization and defectiveness of carbon materials. For Co/Co_2P-NCNFs-H, the I_D/I_G ratio was found to be 1.052, which is significantly higher than that of Co-NCNFs (0.96), Co/Co_2P-NCNFs (0.978), and Co-NCNFs-H (0.982). This increase reflects N and P doping and hollow structure lead to a large number of defects in the carbon nanofibers. These defects play a crucial role in enhancing electrocatalytic activity by increasing the number of active sites and facilitating electron transfer across the carbon network [57]. Moreover, the higher I_D/I_G ratio is consistent with the XPS evidence of increased Co–N bonding and oxygenated functional groups (Figure 3b), suggesting synergistic modulation of the electronic structure through heteroatom doping and interface engineering.

The elemental content in the Co/Co_2P-NCNFs-H sample was determined according to ICP-OES, in which the percentages of the Co and P elements were 4.51 wt% and 0.8 wt%, respectively, as shown in Table S1. Calculated from the thermogravimetric analysis results, it can be concluded that the Co content in Co/Co_2P-NCNFs-H is about 4.46 wt% (Figure 2c), which is basically consistent with the ICP-OES results. The unit loading of elemental Co in this catalytic material was calculated to be about 4.35 mg g^{-1}. The water contact angles of Co/Co_2P-NCNFs-H and Co/Co_2P-NCNFs samples were also tested to evaluate the effect of the hollow carbon nanofiber structure on hydrophilicity and hydrophobicity; the results are shown in Figures S5 and S6. The comparison shows that the Co/Co_2P-NCNFs-H samples with hollow carbon nanofiber structure have better hydrophilicity and aerophobicity, which proves that rational regulation of fiber structure can effectively enhance three-phase interfacial mass transfer in the process of water electrolysis, promote contact between the active sites and the electrolyte solution, and accelerate the escape of bubbles to enhance the catalytic rate of the electrolysis of water.

As obtained by BET nitrogen adsorption-desorption isotherm analysis, the specific surface areas of Co/Co_2P-NCNFs-H (254 m^2 g^{-1}) and Co-NCNFs-H (233 m^2 g^{-1}) were significantly larger than Co-NCNFs (126 m^2 g^{-1}) and Co/Co_2P-NCNFs (144 m^2 g^{-1}), which demonstrated that the hollow structure could significantly enhance the specific surface area of carbon nanofibers to expose more active sites (Figure S7). The specific surface area of Co/Co_2P-NCNFs-H is slightly larger than that of Co-NCNFs-H due to the introduction of defects by P doping in the carbon material. Moreover, the pore sizes of Co-NCNFs and Co/Co_2P-NCNFs are mostly around 1 nm, whereas the hollow carbon nanofiber material Co/Co_2P-NCNFs-H and Co-NCNFs-H samples also have a larger pore distribution, at 2~3 nm (Figure S8), which may be due to the microporous structures left in the inner wall of the hollow fibers after PMMA decomposition. The hollow and porous structures of Co/Co_2P-NCNFs-H and Co-NCNFs-H help to promote rapid mass transfer on the electrode surface and push the reaction forward.

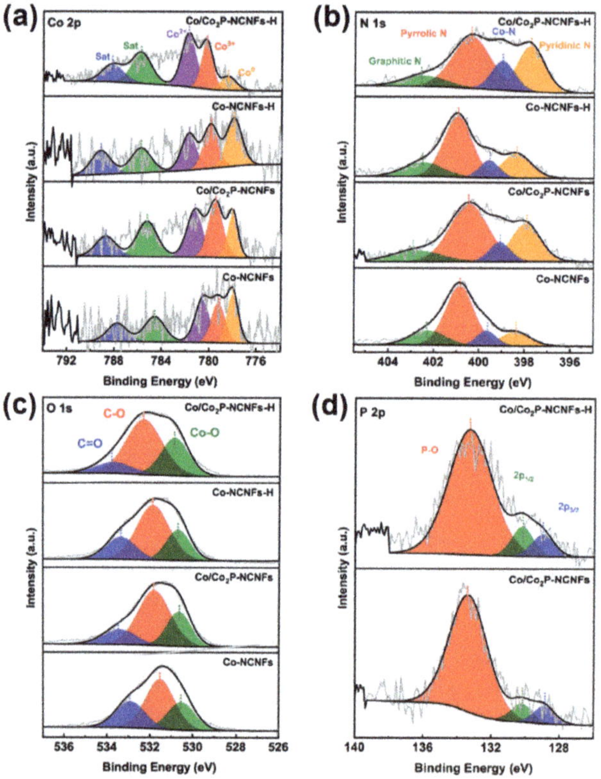

Figure 3. XPS characterization of the samples. (**a**) Co $2p_{3/2}$, (**b**) N 1s, (**c**) O 1s, and (**d**) P 2p for Co-NCNFs, Co/Co_2P-NCNFs, Co-NCNFs-H, and Co/Co_2P-NCNFs-H.

XPS analysis was performed to investigate the surface elemental composition, oxidation states, and chemical structure of the prepared Co-NCNFs, Co/Co_2P-NCNFs, Co-NCNFs-H, and Co/Co_2P-NCNFs-H. As shown in the full-scan XPS spectrum (Figure S9d), the Co/Co_2P-NCNFs-H sample contains Co, C, N, P, and O elements. The corresponding spectra of control samples are presented in Figure S9: all samples exhibit distinct peaks for C 1s, O 1s, and N 1s, while the signals for the Co and P elements appear relatively weak in the full-scan spectra due to their low concentrations. The Co 2p spectrum (Figure 3a) of Co/Co_2P-NCNFs-H reveals a significantly attenuated Co^0 $2p_3/_2$ peak at 778.2 eV compared to other control samples, indicating partial oxidation of metallic cobalt. Simultaneously, the intensity of peaks at 780.6 eV (Co^{3+} $2p_3/_2$) and 782.4 eV (Co^{2+} $2p_3/_2$) increases, accompanied by characteristic satellite features at 786.2 and 792.0 eV, consistent with the formation of Co–O and $Co(OH)_2$ species [53]. These observations are further supported by the O 1s spectrum (Figure 3c), where the dominant peak at 530.1 eV corresponds to metal–oxygen bonds. The N 1s spectrum (Figure 3b) shows four deconvoluted peaks located at 398.4 eV (pyridinic N), 399.8 eV (pyrrolic N), 400.3 eV (Co–N coordination), and 401.1 eV (graphitic N) [58]. Among these, the Co–N component displays a significantly higher intensity in Co/Co_2P-NCNFs-H, indicating that nitrogen atoms are actively involved in coordination with Co species. This suggests a strong electronic interaction between N-doped carbon and cobalt, which may help stabilize active sites and modulate local charge density. The P 2p spectrum (Figure 3d) displays two main peaks at 129.4 eV ($2p_3/_2$) and 130.2 eV ($2p_1/_2$), which are attributed to Co–P bonding, corresponding to the Co_2P diffraction peaks observed in XRD patterns. A broader peak at 133.5 eV corresponds to surface-oxidized phosphorus (P–O) [28]. Notably, the intensity of the Co–P signal is significantly higher in Co/Co_2P-NCNFs-H than in Co/Co_2P-NCNFs, suggesting a higher degree of heterojunction exposure, likely enabled by the hollow and porous architecture. This structure enlarges the interfacial contact among Co, P, and N atoms, reinforcing the both Co-N and Co-P coordination bonds and promoting the stability of active sites [45].

3.3. OER Activity Evaluation of Co/Co_2P-NCNFs-H

The OER catalytic activity and stability of the self-supporting electrodes Co-NCNFs, Co/Co_2P-NCNFs, Co-NCNFs-H, and Co/Co_2P-NCNFs-H were tested in an alkaline electrolyte solution using a three-electrode system. First, the CV curves of the catalysts during the OER process were measured in 1.0 M KOH solution at a scan rate of 5 mV s^{-1}, as shown in Figure 4a. Compared to the control samples, Co/Co_2P-NCNFs-H exhibits the highest OER activity, required an overpotential of only 405.6 mV to achieve a high current density of 100 mA cm^{-2}, which is much lower than Co-NCNFs (586 mV), Co/Co_2P-NCNFs (464.9 mV), and Co-NCNFs-H (488.3 mV) (Figure 4b). The Tafel slopes for Co-NCNFs, Co/Co_2P-NCNFs, Co-NCNFs-H, and Co/Co_2P-NCNFs-H were 145.7 mV dec^{-1}, 91.6 mV dec^{-1}, 110.4 mV dec^{-1}, and 76.2 mV dec^{-1}, respectively (Figure 4c), indicating that Co/Co_2P-NCNFs-H shows the best intrinsic OER activity and reaction kinetics in an alkaline electrolyte solution. The comprehensive performance of Co/Co_2P-NCNFs-H outperformed some of the recently reported transition metal-based OER catalysts (Table 1).

Table 1. OER performance in 1.0 M KOH compared with other transition metal-based catalysts.

Catalysts	Overpotential at 100 mA cm^{-2} (mV)	Tafel Slope (mV dec^{-1})	Ref.
Co/Co_2P-NCNFs-H	405.6	76.2	This work
p-Co_9S_8/NC/CF	398	75	[59]
Ni@N-HCGHF	470	63	[60]
Fe_3C-Co/NC	450	-	[61]
Ni_2P-VP_2/NF	398	56	[62]
Ni/$NiFe_2O_4$	365	86	[63]
Co-ZIF/CDs/CC	401	147	[64]
NiFe LDH/NiS_2/VS_2	384	99	[65]

To compare the ECSA of the prepared samples, CV curves were measured at different scan rates (20, 40, 60, 80, and 100 mV s^{-1}), and the C_{dl} values of the catalysts were calculated (Figure 5a–e). Among them, the C_{dl} values of Co/Co_2P-NCNFs-H and Co-NCNFs-H are similar, at 1.52 mF cm^{-2} and 1.48 mF/cm^{-2}, respectively, which are significantly higher than those of Co/Co_2P-NCNFs (1.01 mF cm^{-2}) and Co-NCNFs (0.9 mF cm^{-2}).

The ECSA was calculated by converting the specific capacitance of a standard electrode with an actual surface area of 1 cm^2 into ECSA, and the results are shown in Figure 5f. The ECSA values of Co-NCNFs, Co/Co_2P-NCNFs, Co-NCNFs-H, and Co/Co_2P-NCNFs-H are 22.5, 25, 37, and 38 $cm^2{}_{ECSA}$, respectively. This demonstrates that the hollow carbon nanofiber structure can significantly increase the electrochemical active surface area of the catalyst, allowing the active sites to better interact with the electrolyte and improving the utilization of active sites. Additionally, the TOF value for Co/Co_2P-NCNFs-H at an overpotential of 450 mV for OER is 27.0 s^{-1} (Figure 4e,f), which is significantly higher than those of the other control samples, further confirming its highest intrinsic OER activity.

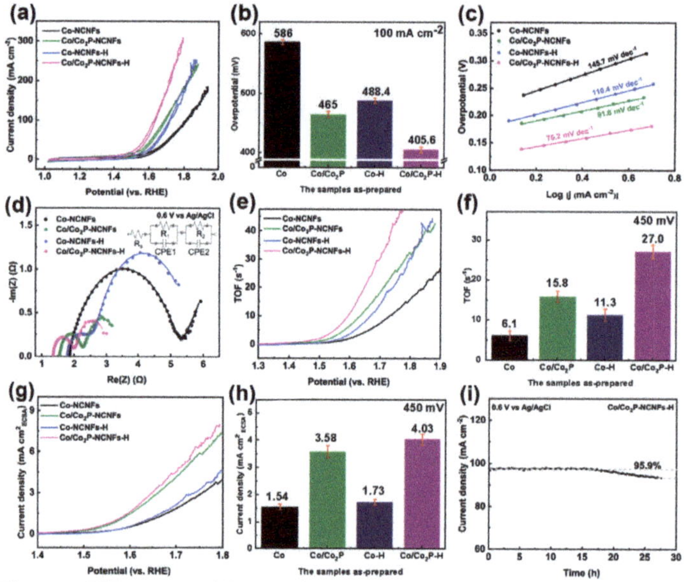

Figure 4. OER activity of the samples. (**a**) OER CV curves, (**b**) OER overpotential at 100 mA cm^{-2}, (**c**) OER Tafel slopes, (**d**) Nyquist plots of OER in 1.0 M KOH, (**e**) TOF curves, (**f**) TOF values at an overpotential of 450 mV, (**g**) OER j_{ECSA} curves, (**h**) j_{ECSA} at an overpotential of 450 mV for Co-NCNFs, Co/Co_2P-NCNFs, Co-NCNFs-H, and Co/Co_2P-NCNFs-H in the OER test in 1.0 M KOH. (**i**) Chronoamperometry curve of Co/Co_2P-NCNFs-H tested at a constant potential of 0.6 V vs. Ag/AgCl in 1.0 M KOH.

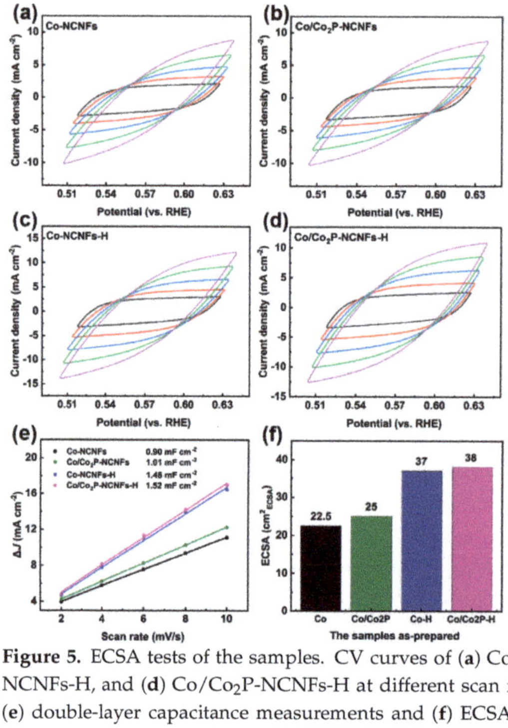

Figure 5. ECSA tests of the samples. CV curves of (**a**) Co-NCNFs, (**b**) Co/Co_2P-NCNFs, (**c**) Co-NCNFs-H, and (**d**) Co/Co_2P-NCNFs-H at different scan rates of 20, 40, 60, 80, and 100 mV s^{-1}; (**e**) double-layer capacitance measurements and (**f**) ECSA of Co-NCNFs, Co/Co_2P-NCNFs, Co-NCNFs-H, and Co/Co_2P-NCNFs-H.

3.4. HER Activity of Co/Co_2P-NCNFs-H

Similarly, in a 1.0 M KOH solution, the HER LSV curves of the catalysts were measured at a scan rate of 5 mV/s, as shown in Figure 6a. Co/Co_2P-NCNFs-H exhibited the best HER activity. Compared to Co-NCNFs (383.8 mV), Co/Co_2P-NCNFs (291.7 mV), and Co-NCNFs-H (314.5 mV), Co/Co_2P-NCNFs-H required the lowest overpotential of only 247.9 mV to reach a current density of 100 mA/cm^2 (Figure 6b). Additionally, the intrinsic HER kinetics of the prepared catalysts were evaluated using the Tafel slope, as shown in Figure 6c. Compared to Co-NCNFs (187.4 mV dec^{-1}), Co/Co_2P-NCNFs (95.4 mV dec^{-1}), and Co-NCNFs-H (117.3 mV dec^{-1}), Co/Co_2P-NCNFs-H exhibited the best intrinsic kinetics, with a Tafel slope of 67.7 mV dec^{-1}. From the electrochemical tests, it can be observed that, compared to Co sites, the Co/Co_2P heterojunction significantly enhances the catalyst's dual-function intrinsic OER and HER activity. Furthermore, the hollow carbon nanofibers can increase the utilization of catalytic active centers by exposing more active sites and provide channels for bubble transfer generated during high-current operation. Therefore, the Co/Co_2P-NCNFs-H and Co-NCNFs-H samples show better electrocatalytic performance compared to Co/Co_2P-NCNFs and Co-NCNFs. Moreover, the TOF value for Co/Co_2P-NCNFs-H at an overpotential of 400 mV for HER is 57.4 s^{-1} (Figure 6e,f), which is also significantly higher than those of the other control samples, further confirming its highest intrinsic HER activity. Many transition metal-based catalysts with excellent performance have been reported in recent years. In comparison, the comprehensive performance of Co/Co_2P-NCNFs-H has surpassed some of them (Table 2).

Table 2. HER performance in 1.0 M KOH compared with other transition metal-based catalysts.

Catalyst	Overpotential at 100 mA cm^{-2} (mV)	Tafel Slope (mV dec^{-1})	Ref.
Co/Co_2P-NCNFs-H	247.9	67.7	This work
$Co(OH)_2$@NiFe/NF	311	74	[66]
Ni_3S_2/$Ni(OH)_2$-5h	360	80.8	[67]
$Ni(OH)_2$/$NiCo_2O_4$	189 (η_{10})	41	[68]
$NiFeCrS_x$/NF	236	67.4	[69]
Co_2P-Ni_3S_2/NF	110	114.2	[70]
ac-$NiCo(OH)_2$/NF	320	90	[71]
F-Co_2P/Fe_2P/IF	151.8	115.01	[72]

Figures 4g,h and 6g,h show the polarization curves of the samples after ECSA normalization, used to verify the catalytic kinetics and intrinsic active sites of the catalysts during the OER and HER processes. By excluding the influence of surface area on the catalyst, the performance curves of Co/Co_2P-NCNFs-H and Co/Co_2P-NCNFs after ECSA normalization are nearly identical in both HER and OER and significantly better than those of Co-NCNFs and Co-NCNFs-H. The results confirm that formation of the Co/Co_2P heterojunction plays an important role in optimizing the intrinsic kinetics of the catalytic electrode, significantly enhancing the intrinsic activity of the catalytic processes for both OER and HER. Among them, Co/Co_2P-NCNFs-H shows the best dual-function intrinsic activity, proving that the hollow structure also contributes to the formation and dispersion of the Co/Co_2P heterojunction.

Furthermore, to compare the resistance of the Co-NCNFs, Co/Co_2P-NCNFs, Co-NCNFs-H, and Co/Co_2P-NCNFs-H samples during the HER and OER processes, EIS measurements were conducted in a 1.0 M KOH solution over a frequency range of 100,000 Hz to 0.1 Hz (Figures 4d and 6d). From the Nyquist plot, the electronic conductivity of Co/Co_2P-NCNFs-H could be enhanced with lower charge-transfer resistance (R_{ct}) of 650 mΩ than Co-NCNFs (3.55 Ω), Co/Co_2P-NCNFs (850 mΩ), and Co-NCNFs-H (3.42 Ω) in the OER. Similarly, in the HER process, Co/Co_2P-NCNFs-H also exhibits the smallest Rct of 270 mΩ compared to Co-NCNFs (520 mΩ), Co/Co_2P-NCNFs (300 mΩ), and Co-NCNFs-H (342 mΩ). This is likely attributed to the activation effect of nitrogen and phosphorus doping in the carbon material and the promotion of charge transfer between interfaces by the Co/Co_2P heterojunction.

The electrochemical stability of Co/Co_2P-NCNFs-H for the OER and the HER was also tested. The self-supporting Co/Co_2P-NCNFs-H electrode exhibited excellent electrochemical stability for both the OER and the HER in a 1.0 M KOH solution. It was able to stably operate for nearly 20 h at 0.6 V (vs. Ag/AgCl) (Figure 4i) and for nearly 250 h at −1.5 V (vs. Ag/AgCl) without significant performance degradation (Figure 6i). This demonstrates that doping of the N and P elements in the carbon material successfully anchors the Co/Co_2P heterojunction active sites, effectively preventing aggregation and detachment of metal elements during long-term electrochemical testing, thus improving stability. The excellent mechanical strength and stable three-dimensional structure of the hollow carbon nanofiber structure also ensure long-term stable operation of the electrode's active sites.

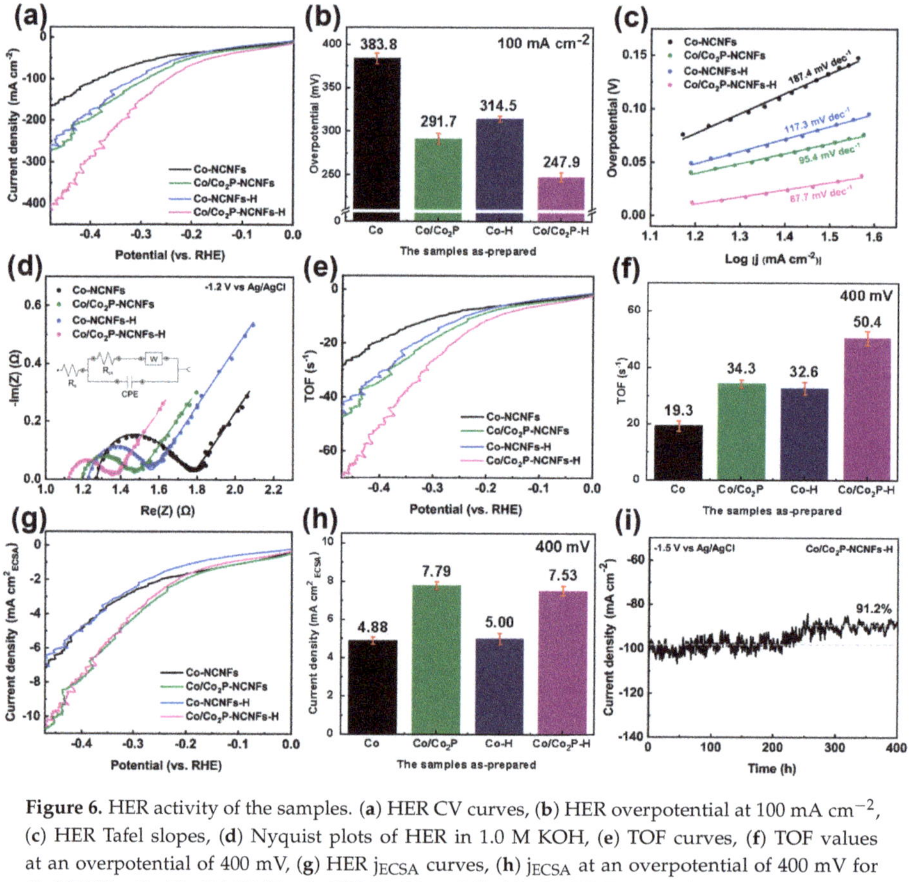

Figure 6. HER activity of the samples. (**a**) HER CV curves, (**b**) HER overpotential at 100 mA cm^{-2}, (**c**) HER Tafel slopes, (**d**) Nyquist plots of HER in 1.0 M KOH, (**e**) TOF curves, (**f**) TOF values at an overpotential of 400 mV, (**g**) HER j_{ECSA} curves, (**h**) j_{ECSA} at an overpotential of 400 mV for Co-NCNFs, Co/Co_2P-NCNFs, Co-NCNFs-H, and Co/Co_2P-NCNFs-H in the HER test in 1.0 M KOH. (i) Chronoamperometry curve of Co/Co_2P-NCNFs-H tested at a constant potential of −1.5 V (vs. Ag/AgCl).

3.5. Characterization of Co/Co_2P-NCNFs-H After OER and HER Stability Tests

The microstructure of the Co/Co_2P-NCNFs-H samples after stability testing, denoted as Co/Co_2P-NCNFs-H-After (OER) and Co/Co_2P-NCNFs-H-After (HER), was characterized by SEM and TEM. After a long electrochemical oxidation process (Figures 7a and S10), the three-dimensional structure and hollow channels of the carbon nanofibers were well preserved, with no significant change in fiber diameter. However, partial aggregation of the metal particles encapsulated in the carbon shell was observed on the fibers, which may be the cause of the catalyst's performance degradation. After more than 200 h of HER testing (Figures 7b and S11), the uniformity and continuity of the Co/Co_2P-NCNFs-H-After (HER) fibers, as well as the fiber thickness, were well maintained, and no significant aggregation of metal elements was found on the fibers. These results collectively demonstrate that the Co/Co_2P-NCNFs-H sample has good microstructural stability and mechanical strength during electrochemical testing.

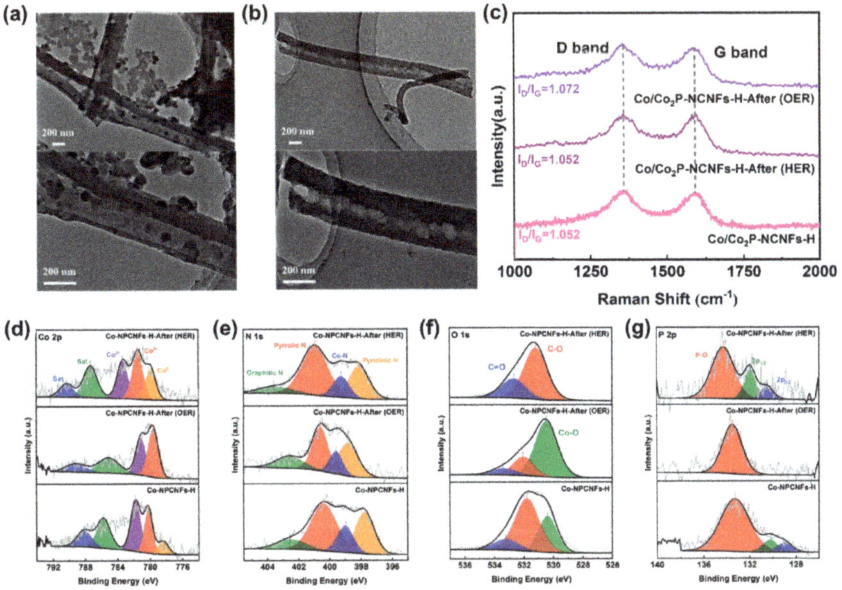

Figure 7. Structural stability during electrocatalysis. (**a**) TEM images of Co/Co_2P-NCNFs-H-After (OER). (**b**) TEM images of Co/Co_2P-NCNFs-H-After (HER). (**c**) Raman, (**d**) Co 2p XPS, (**e**) N 1s XPS, (**f**) O 1s XPS, and (**g**) P 2p XPS spectra of Co/Co_2P-NCNFs-H-After (OER) and Co/Co_2P-NCNFs-H-After (HER).

Raman analysis was also performed on the post-reaction samples, as shown in Figure 7c. After the HER and OER reactions, the I_D/I_G ratio of the samples showed no significant changes, indicating excellent stability of the carbon material during the electrocatalytic process. Figure 7d–g presents the XPS characterization of the Co/Co_2P-NCNFs-H-After (OER) and Co/Co_2P-NCNFs-H-After (HER) samples. The Co 2p spectrum of Co/Co_2P-NCNFs-H-After (OER) shows that the Co^0 peak disappears and is oxidized to a higher Co^{2+}/Co^{3+} oxidation state, with a significant increase in the peak area of the M-O coordination in the O 1s spectrum. Additionally, only the P-O coordination feature peak remains in the P 2p spectrum, while the characteristic peaks of P bonding with Co in the $2p_{1/2}$ and $2p_{3/2}$ orbitals vanish. This could be due to the strong oxidation process, which causes the Co element to partially leach out, weakening the coordination strength between Co and P. For the Co/Co_2P-NCNFs-H-After (HER) sample, the Co 2p spectrum shows an increased proportion of the Co^0 peak, and the Co-O coordination peak in the O 1s spectrum disappears, confirming that the metal elements are reduced to a lower oxidation state. The P 2p spectrum shows a decrease in the intensity of the P-O coordination feature peak, indicating a decrease in the overall oxidation state of the sample after the HER.

Future work should focus on enhancing the catalyst's stability, particularly for the oxygen evolution reaction (OER). While the material demonstrated excellent durability for the hydrogen evolution reaction (HER) for over 250 h, its OER stability was shorter, due to metal particle aggregation. Strategies like surface passivation or doping with oxophilic elements could mitigate this oxidative degradation. A crucial next step is to evaluate the electrode in a practical two-electrode electrolyzer to assess its performance under industrially relevant conditions.

To deepen mechanistic understanding, in situ and operando spectroscopy combined with density functional theory (DFT) calculations should be employed. This approach can elucidate the precise nature of the active sites at the Co/Co_2P heterojunction and clarify the reaction pathways for both the HER and the OER. Finally, the versatility of the continuous coaxial co-spinning method should be leveraged to synthesize other catalysts, such as different transition metal phosphides or sulfides (e.g., Ni, Fe, Mo), to further tune catalytic activity. A thorough techno-economic analysis is also warranted to confirm the commercial viability of this scalable synthesis strategy for sustainable hydrogen production.

4. Conclusions

In conclusion, this work successfully demonstrates the importance and effectiveness of synergistically combining structural and electronic design principles to create a highly efficient and durable bifunctional electrocatalyst for overall water splitting. A self-supporting N-doped hollow carbon nanofiber electrode anchored with Co/Co_2P heterojunctions (Co/Co_2P-NCNFs-H) was synthesized via continuous coaxial electrospinning. The significance of this catalyst is rooted in its unique architecture, where the hollow and porous structure provides abundant active sites and facilitates mass/gas transport, while the Co/Co_2P heterojunction interface optimizes the electronic configuration to boost the intrinsic catalytic activity for both the HER and the OER. The material demonstrated exceptional bifunctional electrocatalytic activity for overall water splitting in 1.0 M KOH, achieving a current density of 100 mA cm^{-2} at low overpotentials of 405.6 mV for the OER and 247.9 mV for the HER, alongside outstanding long-term stability (maintaining operation for 20 h at 0.6 V vs. Ag/AgCl for the OER and 250 h at −1.5 V vs. Ag/AgCl for the HER without significant degradation). Structural analyses revealed a uniform hollow nanofiber architecture with robust mechanical stability, while BET measurements confirmed a high specific surface area (254 m^2 g^{-1}) and porous structure, facilitating abundant active site exposure. Hydrophilic/superaerophobic properties validated by contact angle tests enhanced electrolyte infiltration and gas dissipation at the triple-phase interface. XRD and XPS confirmed the coexistence of metallic Co and Co_2P, forming heterojunction interfaces that optimized electron transfer kinetics. Comparative studies with control samples (Co-NCNFs, Co-NCNFs-H, and Co/Co_2P-NCNFs) highlighted the synergistic interplay between the hollow structure and heterojunction in boosting catalytic performance. This work provides new ideas for the rational design and preparation of hollow carbon nanofiber catalytic materials with large specific surface area, fast mass transfer channels, and uniform dispersion of active sites.

Supplementary Materials: The following supporting information can be downloaded at: https://www.mdpi.com/article/10.3390/coatings15070772/s1, Figure S1. Schematic illustration of the overall synthesis of the Co/Co_2P-NCNFs-H. Figure S2. HRTEM image of nanoparticle on Co/Co_2P-NCNFs-H. Figure S3. (a) SEM images, (b) TEM image and (c) HRTEM image of nanoparticles on

Co/Co_2P-NCNFs. Figure S4. HAADF-STEM image and EDS elemental maps of the Co/Co_2P-NCNFs. Figure S5. Droplet (a) and bubble (b,c) contact angles (CA) of Co/Co_2P-NCNFs-H. Figure S6. Droplet (a) and bubble (b,c) contact angles (CA) of Co/Co_2P-NCNFs. Figure S7. N_2 adsorption-desorption isotherms of Co/Co_2P-NCNFs-H. Figure S8. Pore size distributions of Co-NCNFs, Co/Co_2P-NCNFs, Co-NCNFs-H and Co/Co_2P-NCNFs-H. Figure S9. Full scan XPS spectra of Co-NCNFs, Co/Co_2P-NCNFs, Co-NCNFs-H and Co/Co_2P-NCNFs-H. Figure S10. SEM image of Co/Co_2P-NCNFs-H-After (OER). Figure S11. SEM image of Co/Co_2P-NCNFs-H-After (HER). Figure S12. The diagram of overall water splitting on Co/Co_2P-NCNFs-H [2,73]. Table S1. The ICP analysis results of Co/Co_2P-NCNFs-H. Table S2. The FHWM and binding energy used during XPS deconvolution.

Author Contributions: Conceptualization, R.D.; Formal analysis, R.D., J.D. and J.F.; Investigation, R.D., J.D. and J.F.; Writing—original draft, R.D. and J.D.; Writing—review & editing, L.Z.; Supervision, L.Z.; Project administration, L.Z. All authors have read and agreed to the published version of the manuscript.

Funding: This work was supported by research funding provided by National Natural Science Foundation of China (Grant Nos. 22378119 and 22208092).

Institutional Review Board Statement: Not applicable.

Informed Consent Statement: Not applicable.

Data Availability Statement: The data that support the findings of this study are available from the corresponding author upon reasonable request.

Conflicts of Interest: The authors declare no conflicts of interest.

References

1. Küspert, S.; Campbell, I.E.; Zeng, Z.; Balaghi, S.E.; Ortlieb, N.; Thomann, R.; Knäbbeler-Buß, M.; Allen, C.S.; Mohney, S.E.; Fischer, A. Ultrasmall and Highly Dispersed Pt Entities Deposited on Mesoporous N-doped Carbon Nanospheres by Pulsed CVD for Improved HER. *Small* **2024**, *20*, 2311260. [CrossRef]
2. Zhang, W.; Liu, M.; Gu, X.; Shi, Y.; Deng, Z.; Cai, N. Water Electrolysis toward Elevated Temperature: Advances, Challenges and Frontiers. *Chem. Rev.* **2023**, *123*, 7119–7192. [CrossRef]
3. Jiao, Y.; Zheng, Y.; Jaroniec, M.; Qiao, S.Z. Design of electrocatalysts for oxygen- and hydrogen-involving energy conversion reactions. *Chem. Soc. Rev.* **2015**, *44*, 2060–2086. [CrossRef]
4. Kou, Z.; Liu, Y.; Cui, W.; Yang, B.; Li, Z.; Rodriguez, R.D.; Zhang, Q.; Dong, C.-L.; Sang, X.; Lei, L.; et al. Electronic structure optimization of metal–phthalocyanine via confining atomic Ru for all-pH hydrogen evolution. *Energy Environ. Sci.* **2024**, *17*, 1540–1548. [CrossRef]
5. Zainal, B.S.; Ker, P.J.; Mohamed, H.; Ong, H.C.; Fattah, I.M.R.; Rahman, S.M.A.; Nghiem, L.D.; Mahlia, T.M.I. Recent advancement and assessment of green hydrogen production technologies. *Renew. Sustain. Energy Rev.* **2024**, *189*, 113941. [CrossRef]
6. Li, L.; Wang, P.; Shao, Q.; Huang, X. Recent progress in advanced electrocatalyst design for acidic oxygen evolution reaction. *Adv. Mater.* **2021**, *33*, 2004243. [CrossRef]
7. Quan, L.; Jiang, H.; Mei, G.; Sun, Y.; You, B. Bifunctional Electrocatalysts for Overall and Hybrid Water Splitting. *Chem. Rev.* **2024**, *124*, 3694–3812. [CrossRef]
8. Deng, L.; Hung, S.-F.; Lin, Z.-Y.; Zhang, Y.; Zhang, C.; Hao, Y.; Liu, S.; Kuo, C.-H.; Chen, H.-Y.; Peng, J.; et al. Valence Oscillation of Ru Active Sites for Efficient and Robust Acidic Water Oxidation. *Adv. Mater.* **2023**, *35*, 2305939. [CrossRef]
9. Paudel, D.R.; Pan, U.N.; Ghising, R.B.; Dhakal, P.P.; Dinh, V.A.; Wang, H.; Kim, N.H.; Lee, J.H. Interface modulation induced by the 1T Co-WS_2 shell nanosheet layer at the metallic $NiTe_2$/Ni core–nanoskeleton: Glib electrode-kinetics for HER, OER, and ORR. *Nano Energy* **2022**, *102*, 107712. [CrossRef]
10. Zhao, S.; Hu, F.; Yin, L.; Li, L.; Peng, S. Manipulating electron redistribution induced by asymmetric coordination for electrocatalytic water oxidation at a high current density. *Sci. Bull.* **2023**, *68*, 1389–1398. [CrossRef]
11. Song, Z.; Zhu, Y.-N.; Liu, H.; Banis, M.N.; Zhang, L.; Li, J.; Doyle-Davis, K.; Li, R.; Sham, T.-K.; Yang, L.; et al. Engineering the Low Coordinated Pt Single Atom to Achieve the Superior Electrocatalytic Performance toward Oxygen Reduction. *Small* **2020**, *16*, 2003096. [CrossRef]
12. Tan, L.; Wang, H.; Qi, C.; Peng, X.; Pan, X.; Wu, X.; Wang, Z.; Ye, L.; Xiao, Q.; Luo, W.; et al. Regulating Pt electronic properties on NiFe layered double hydroxide interface for highly efficient alkaline water splitting. *Appl. Catal. B Environ.* **2024**, *342*, 123352. [CrossRef]

13. Ngo, Q.P.; Nguyen, T.T.; Le, Q.T.T.; Lee, J.H.; Kim, N.H. Unveiling the Synergistic Effect of Atomic Iridium Modulated Zirconium-Doped Pure Phase Cobalt Phosphide for Robust Anion-Exchange Membrane Water Electrolyzer. *Adv. Energy Mater.* **2023**, *13*, 2301841. [CrossRef]
14. Chen, Y.; Liu, Y.; Zhai, W.; Liu, H.; Sakthivel, T.; Guo, S.; Dai, Z. Metastabilizing the Ruthenium Clusters by Interfacial Oxygen Vacancies for Boosted Water Splitting Electrocatalysis. *Adv. Energy Mater.* **2024**, *14*, 2400059. [CrossRef]
15. Yu, L.; Zhu, Q.; Song, S.; McElhenny, B.; Wang, D.; Wu, C.; Qin, Z.; Bao, J.; Yu, Y.; Chen, S.; et al. Non-noble metal-nitride based electrocatalysts for high-performance alkaline seawater electrolysis. *Nat. Commun.* **2019**, *10*, 5106. [CrossRef]
16. Qian, J.; Liu, X.; Zhong, C.; Xu, G.; Li, H.; Zhou, W.; You, B.; Wang, F.; Gao, D.; Chao, D. Enhanced stability and narrowed D-band gap of Ce-doped Co3O4 for rechargeable aqueous Zn-air battery. *Adv. Funct. Mater.* **2023**, *33*, 2212021. [CrossRef]
17. Yu, M.; Budiyanto, E.; Tüysüz, H. Principles of Water Electrolysis and Recent Progress in Cobalt-, Nickel-, and Iron-Based Oxides for the Oxygen Evolution Reaction. *Angew. Chem. Int. Ed.* **2022**, *61*, e202103824. [CrossRef]
18. Browne, M.P.; Sofer, Z.; Pumera, M. Layered and two dimensional metal oxides for electrochemical energy conversion. *Energy Environ. Sci.* **2019**, *12*, 41–58. [CrossRef]
19. Ren, J.-T.; Yuan, G.-G.; Weng, C.-C.; Chen, L.; Yuan, Z.-Y. Uniquely integrated Fe-doped Ni(OH)2 nanosheets for highly efficient oxygen and hydrogen evolution reactions. *Nanoscale* **2018**, *10*, 10620–10628. [CrossRef]
20. Qiao, L.; Zhu, A.; Zeng, W.; Dong, R.; Tan, P.; Ding, Z.; Gao, P.; Wang, S.; Pan, J. Achieving electronic structure reconfiguration in metallic carbides for robust electrochemical water splitting. *J. Mater. Chem. A* **2020**, *8*, 2453–2462. [CrossRef]
21. Chen, P.; Ye, J.; Wang, H.; Ouyang, L.; Zhu, M. Recent progress of transition metal carbides/nitrides for electrocatalytic water splitting. *J. Alloys Compd.* **2021**, *883*, 160833. [CrossRef]
22. Weng, C.-C.; Ren, J.-T.; Yuan, Z.-Y. Transition Metal Phosphide-Based Materials for Efficient Electrochemical Hydrogen Evolution: A Critical Review. *ChemSusChem* **2020**, *13*, 3357–3375. [CrossRef]
23. Fang, Y.; Luan, D.; Lou, X.W. Recent Advances on Mixed Metal Sulfides for Advanced Sodium-Ion Batteries. *Adv. Mater.* **2020**, *32*, 2002976. [CrossRef]
24. Wang, H.Y.; Ren, J.T.; Wang, L.; Sun, M.L.; Yang, H.M.; Lv, X.W.; Yuan, Z.Y. Synergistically enhanced activity and stability of bifunctional nickel phosphide/sulfide heterointerface electrodes for direct alkaline seawater electrolysis. *J. Energy Chem.* **2022**, *75*, 66–73. [CrossRef]
25. Gao, W.; Wu, Y.J.; Wan, X.H.; Gao, J.; Wen, D. Engineering the electronic structure of FeP with rare earth elements to enhance the electrocatalytic hydrogen evolution performance. *J. Mater. Chem. A* **2023**, *11*, 18126–18134. [CrossRef]
26. Li, C.; Hong, W.T.; Cai, Q.; Jian, C.Y. Directional Construction of a $1T_{0.63}$-$MoSe_2$@MoP Multiphase-Interface Catalyst for Highly Efficient Alkaline Hydrogen Evolution. *ACS Appl. Mater. Interfaces* **2022**, *14*, 30683–30691. [CrossRef]
27. Song, M.; Wang, J.B.; Zhang, X.; Guo, J.X. CoP@NC nanorod array for efficient alkaline water and alkaline seawater electrolysis. *J. Alloys Compd.* **2025**, *1018*, 179248. [CrossRef]
28. Zhang, C.X.; Li, F.T.; Wu, D.; Guo, Q.M.; Liu, Z.N.; Wang, Z.K.; Kang, Z.X.; Fan, L.L.; Sun, D.F. Oxygen-coordinated MOF membrane facilitated construction of supported Co2P/CoP@C heterostructures for water electrolysis. *Inorg. Chem. Front.* **2025**, *12*, 2254–2265. [CrossRef]
29. Ipadeola, A.K.; Sliem, M.H.; Mwonga, P.; Ozoemena, K.I.; Abdullah, A.M. Rational construction of N-containing carbon sheets atomically doped NiP-CoP nanohybrid electrocatalysts for enhanced green hydrogen and oxygen production. *Electrochim. Acta* **2024**, *508*, 145236. [CrossRef]
30. Huo, J.J.; Ming, Y.; Huang, X.L.; Ge, R.Y.; Li, S.; Zheng, R.K.; Cairney, J.; Dou, S.X.; Fei, B.; Li, W.X. Arrayed metal phosphide heterostructure by Fe doping for robust overall water splitting. *J. Colloid Interface Sci.* **2025**, *678*, 669–681. [CrossRef]
31. Li, W.; Liu, J.; Guo, P.; Li, H.; Fei, B.; Guo, Y.; Pan, H.; Sun, D.; Fang, F.; Wu, R. Co/CoP Heterojunction on Hierarchically Ordered Porous Carbon as a Highly Efficient Electrocatalyst for Hydrogen and Oxygen Evolution. *Adv. Energy Mater.* **2021**, *11*, 2102134. [CrossRef]
32. Zhang, Y.; Li, Z.Y.; He, S.Q.; Qiao, Y.X.; Yuan, A.H.; Wu, J.C.; Zhou, H. Interfacial engineering of heterostructured CoP/FeP nanoflakes as bifunctional electrocatalyts toward alkaline water splitting. *J. Colloid Interface Sci.* **2025**, *679*, 20–29. [CrossRef]
33. Li, J.W.; Hu, Y.Z.; Huang, X.; Zhu, Y.; Wang, D.L. Bimetallic Phosphide Heterostructure Coupled with Ultrathin Carbon Layer Boosting Overall Alkaline Water and Seawater Splitting. *Small* **2023**, *19*, e2206533. [CrossRef]
34. Zhao, D.; Zhu, Y.; Wu, Q.; Zhou, W.; Dan, J.; Zhu, H.; Lei, W.; Ma, L.-J.; Li, L. Low-loading Ir decorated cobalt encapsulated N-doped carbon nanotubes/porous carbon sheet implements efficient hydrogen/oxygen trifunctional electrocatalysis. *Chem. Eng. J.* **2022**, *430*, 132825. [CrossRef]
35. Yu, B.; Liu, J.-H.; Guo, S.; Huang, G.; Zhang, S.; Chen, S.; Li, X.; Wang, Y.; Lv, L.-P. Densely populated tiny RuO2 crystallites supported by hierarchically porous carbon for full acidic water splitting. *Mater. Horiz.* **2023**, *10*, 4589–4596. [CrossRef]
36. Jiang, Y.; Yang, L.; Sun, T.; Zhao, J.; Lyu, Z.; Zhuo, O.; Wang, X.; Wu, Q.; Ma, J.; Hu, Z. Significant Contribution of Intrinsic Carbon Defects to Oxygen Reduction Activity. *ACS Catal.* **2015**, *5*, 6707–6712. [CrossRef]

37. Wang, H.-F.; Tang, C.; Zhang, Q. Template growth of nitrogen-doped mesoporous graphene on metal oxides and its use as a metal-free bifunctional electrocatalyst for oxygen reduction and evolution reactions. *Catal. Today* **2018**, *301*, 25–31. [CrossRef]
38. Zhang, H.; Yang, T.; Zhou, H.; Wang, Y.; Yang, X.; Liang, W.; Wu, D.; Yuan, P.; Yu, T.; He, M.; et al. Activating Fe activity and improving Ni activity via C3N4 substrate in alkaline oxygen evolution catalyzed by Ni-Fe phosphide. *Appl. Catal. B Environ.* **2024**, *342*, 123391. [CrossRef]
39. Zhang, J.; Huang, S.S.; Ning, P.; Xin, P.J.; Chen, Z.W.; Wang, Q.; Uvdal, K.; Hu, Z.J. Nested hollow architectures of nitrogen-doped carbon-decorated Fe, Co, Ni-based phosphides for boosting water and urea electrolysis. *Nano Res.* **2022**, *15*, 1916–1925. [CrossRef]
40. Zhou, T.; Zhou, Y.; Ma, R.; Zhou, Z.; Liu, G.; Liu, Q.; Zhu, Y.; Wang, J. Nitrogen-doped hollow mesoporous carbon spheres as a highly active and stable metal-free electrocatalyst for oxygen reduction. *Carbon* **2017**, *114*, 177–186. [CrossRef]
41. Zhu, R.L.; Yu, X.L.; Li, W.C.; Li, M.; Bo, X.J.; Gan, G.Y. Cobalt nanoparticles-embedded porous carbon nanocages uniformly dispersed hollow carbon fibers as the accelerated electrocatalysts toward water splitting. *J. Alloys Compd.* **2023**, *947*, 169488. [CrossRef]
42. Zhang, J.; Li, X.T.; Li, T.; Zheng, J.L.; Yang, W.; Xiao, Y.Q. Ruthenium nanoparticles incorporated hollow carbon nanoshells for improved hydrogen evolution reaction. *Int. J. Hydrogen Energy* **2024**, *51*, 1033–1041. [CrossRef]
43. Guan, Z.; Li, J.; Li, S.; Wang, K.; Lei, L.; Wang, Y.; Zhuang, L.; Xu, Z. The transient covering of iridium species with ultrathin carbon shells via Joule-heating for robust acidic water oxidation. *Mater. Chem. Front.* **2024**, *8*, 824–835. [CrossRef]
44. Poudel, M.B.; Logeshwaran, N.; Kim, A.R.; Karthikeyan, S.C.; Vijayapradeep, S.; Yoo, D.J. Integrated core-shell assembly of Ni_3S_2 nanowires and CoMoP nanosheets as highly efficient bifunctional electrocatalysts for overall water splitting. *J. Alloys Compd.* **2023**, *960*, 170678. [CrossRef]
45. Tong, J.; Li, Y.; Bo, L.; Li, W.; Li, T.; Zhang, Q.; Kong, D.; Wang, H.; Li, C. CoP/N-doped carbon hollow spheres anchored on electrospinning core–shell N-doped carbon nanofibers as efficient electrocatalysts for water splitting. *ACS Sustain. Chem. Eng.* **2019**, *7*, 17432–17442. [CrossRef]
46. Patil, S.A.; Khot, A.C.; Chavan, V.D.; Rabani, I.; Kim, D.K.; Jung, J.; Im, H.; Shrestha, N.K. Electrostatically robust CoFeOF nanosheet against chloride for green-H_2 production in alkaline seawater electrolysis. *Chem. Eng. J.* **2024**, *480*, 146545. [CrossRef]
47. Tran, V.A.; Do, H.H.; Le, V.T.; Vasseghian, Y.; Vo, V.; Ahn, S.H.; Kim, S.Y.; Lee, S.W. Metal-organic-framework-derived metals and metal compounds as electrocatalysts for oxygen evolution reaction: A review. *Int. J. Hydrogen Energy* **2022**, *47*, 19590–19608. [CrossRef]
48. Zhao, X.L.; Yong, X.H.; Ji, Q.Z.; Yang, Z.H.; Song, Y.; Tian, T.Y.; Chen, T.; Yang, Z.G.; Xu, L.X.; Shen, X.; et al. Synthesis of all-biomass-derived carbon nanofibers for dual-functional filtration membranes and oxygen evolution reaction electrocatalysts. *J. Alloys Compd.* **2022**, *918*, 165600. [CrossRef]
49. Poudel, M.B.; Kim, A.A.; Lohani, P.C.; Yoo, D.J.; Kim, H.J. Assembling zinc cobalt hydroxide/ternary sulfides heterostructure and iron oxide nanorods on three-dimensional hollow porous carbon nanofiber as high energy density hybrid supercapacitor. *J. Energy Storage* **2023**, *60*, 106713. [CrossRef]
50. Peng, L.C.; Wang, J.A.; Guo, S.Q.; Li, C.J. Exploratory construction of Co/Co_3O_4-Ni/NiO heterointerface modified macroporous interconnected hollow carbon nanofibers towards efficient and flexible electrocatalysis. *Chem. Eng. J.* **2022**, *450*, 138252. [CrossRef]
51. Papkov, D.; Delpouve, N.; Delbreilh, L.; Araujo, S.; Stockdale, T.; Mamedov, S.; Maleckis, K.; Zou, Y.; Andalib, M.N.; Dargent, E. Quantifying polymer chain orientation in strong and tough nanofibers with low crystallinity: Toward next generation nanostructured superfibers. *ACS Nano* **2019**, *13*, 4893–4927. [CrossRef]
52. Wang, X.W.; Su, T.T.; Lu, Z.H.; Yu, L.; Sha, N.; Lv, C.M.; Xie, Y.; Ye, K. Morphological engineering of monodispersed Co2P nanocrystals for efficient alkaline water and seawater splitting. *J. Colloid Interface Sci.* **2025**, *691*, 137389. [CrossRef] [PubMed]
53. Ye, L.; Chen, W.; Jiang, Z.J.; Jiang, Z. Co/CoO heterojunction rich in oxygen vacancies introduced by O_2 plasma embedded in mesoporous walls of carbon nanoboxes covered with carbon nanotubes for rechargeable zinc–air battery. *Carbon Energy* **2024**, *6*, e457. [CrossRef]
54. Kandel, M.R.; Pan, U.N.; Dhakal, P.P.; Ghising, R.B.; Nguyen, T.T.; Zhao, J.; Kim, N.H.; Lee, J.H. Unique heterointerface engineering of Ni_2P- MnP nanosheets coupled Co_2P nanoflowers as hierarchical dual-functional electrocatalyst for highly proficient overall water-splitting. *Appl. Catal. B Environ.* **2023**, *331*, 122680. [CrossRef]
55. Peng, X.; Jin, X.; Gao, B.; Liu, Z.; Chu, P.K. Strategies to improve cobalt-based electrocatalysts for electrochemical water splitting. *J. Catal.* **2021**, *398*, 54–66. [CrossRef]
56. Xia, C.; Huang, L.; Yan, D.; Douka, A.I.; Guo, W.; Qi, K.; Xia, B.Y. Electrospinning synthesis of self-standing cobalt/nanocarbon hybrid membrane for long-life rechargeable zinc–air batteries. *Adv. Funct. Mater.* **2021**, *31*, 2105021. [CrossRef]
57. Jia, Y.; Jiang, K.; Wang, H.; Yao, X. The Role of Defect Sites in Nanomaterials for Electrocatalytic Energy Conversion. *Chem* **2019**, *5*, 1371–1397. [CrossRef]
58. Ahn, C.H.; Yang, W.S.; Kim, J.J.; Priyanga, G.S.; Thomas, T.; Deshpande, N.G.; Lee, H.S.; Cho, H.K. Design of hydrangea-type Co/Mo bimetal MOFs and MOF-derived Co/Mo2C embedded carbon composites for highly efficient oxygen evolution reaction. *Chem. Eng. J.* **2022**, *435*, 134815. [CrossRef]

59. Zhao, Y.G.; Li, T.; Wang, Q.; Ai, Y.Y.; Hou, R.H.; Habib, A.; Shao, G.S.; Wang, F.; Zhang, P. A parallel array structured cobalt sulfide/nitrogen doped carbon nanocage/carbon fiber composite based on microfluidic spinning technology: A novel design to boost overall water splitting. *J. Mater. Chem. A* **2024**, *12*, 23872–23879. [CrossRef]
60. Yan, L.; Xu, Y.; Chen, P.; Zhang, S.; Jiang, H.; Yang, L.; Wang, Y.; Zhang, L.; Shen, J.; Zhao, X. Heterostructure Films: A Freestanding 3D Heterostructure Film Stitched by MOF-Derived Carbon Nanotube Microsphere Superstructure and Reduced Graphene Oxide Sheets: A Superior Multifunctional Electrode for Overall Water Splitting and Zn–Air Batteries (Adv. Mater. 48/2020). *Adv. Mater.* **2020**, *32*, 2070362.
61. Yang, C.C.; Zai, S.F.; Zhou, Y.T.; Du, L.; Jiang, Q. Fe_3C-Co nanoparticles encapsulated in a hierarchical structure of N-doped carbon as a multifunctional electrocatalyst for ORR, OER, and HER. *Adv. Funct. Mater.* **2019**, *29*, 1901949. [CrossRef]
62. Yan, H.; Xie, Y.; Wu, A.; Cai, Z.; Wang, L.; Tian, C.; Zhang, X.; Fu, H. Anion-modulated HER and OER activities of 3D Ni–V-based interstitial compound heterojunctions for high-efficiency and stable overall water splitting. *Adv. Mater.* **2019**, *31*, 1901174. [CrossRef] [PubMed]
63. Kubisztal, J.; Kubisztal, M. Pressed Ni/MFe_2O_4 (M = Ni, Co) powder compacts for application as bifunctional, high-performance electrodes in electrochemical water splitting. *Int. J. Hydrogen Energy* **2024**, *56*, 912–923. [CrossRef]
64. Hong, Q.; Wang, Y.; Wang, R.; Chen, Z.; Yang, H.; Yu, K.; Liu, Y.; Huang, H.; Kang, Z.; Menezes, P.W. In Situ Coupling of Carbon Dots with Co-ZIF Nanoarrays Enabling Highly Efficient Oxygen Evolution Electrocatalysis. *Small* **2023**, *19*, 2206723. [CrossRef] [PubMed]
65. Wang, T.; Zhang, X.; Yu, X.; Li, J.; Wang, K.; Niu, J. Interfacial Interaction in NiFe LDH/$NiS_{(2)}$/$VS_{(2)}$ for Enhanced Electrocatalytic Water Splitting. *Molecules* **2024**, *29*, 951. [CrossRef] [PubMed]
66. Wang, H.; Yan, Y.; Zhang, W.; Sun, S.; Yao, S. Preparation of $Co(OH)_2$@NiFe/NF bifunctional electrocatalyst by electrodeposition for efficient water splitting. *J. Solid State Chem.* **2023**, *323*, 124048. [CrossRef]
67. Dai, X.; Tang, Z.; Yan, X.; Tao, S.; Wang, S.; Liu, Y.; Cao, J.; Deng, X.; Bo, X. In-Situ Sulfuration of $Ni(OH)_2$ to Heterostructured Ni_3S_2/$Ni(OH)_2$@Ni Catalyst for Efficient Water Splitting. *Chem. Asian J.* **2025**, *20*, e202401190. [CrossRef]
68. Sang, Y.; Cao, X.; Wang, L.; Ding, G.; Wang, Y.; Yu, D.; Hao, Y.; Li, L.; Peng, S. Facile synthesis of three-dimensional spherical $Ni(OH)_2$/$NiCo_2O_4$ heterojunctions as efficient bifunctional electrocatalysts for water splitting. *Int. J. Hydrogen Energy* **2020**, *45*, 30601–30610. [CrossRef]
69. He, Y.; Yang, X.; Jiang, M.; Liu, F.; Zhang, J.; Li, H.; Cui, L.; Xu, J.; Ji, X.; Liu, J. Cr-doped NiFe sulfides nanoplate array: Highly efficient and robust bifunctional electrocatalyst for the overall water splitting and seawater electrolysis. *J. Colloid Interface Sci.* **2025**, *680*, 1079–1089. [CrossRef]
70. Li, H.; Gao, X.; Li, G. Construction of Co_2P-Ni_3S_2/NF Heterogeneous Structural Hollow Nanowires as Bifunctional Electrocatalysts for Efficient Overall Water Splitting. *Small* **2023**, *19*, 2304081. [CrossRef]
71. Ju, S.; Liu, Y.; Pei, M.; Shuai, Y.; Zhai, Z.; Yan, W.; Wang, Y.-J.; Zhang, J. Amorphization-induced abundant coordinatively unsaturated Ni active sites in $NiCo(OH)_2$ for boosting catalytic OER and HER activities at high current densities for water-electrolysis. *J. Colloid Interface Sci.* **2024**, *653*, 1704–1714. [CrossRef] [PubMed]
72. Zhang, X.-Y.; Zhu, Y.-R.; Chen, Y.; Dou, S.-Y.; Chen, X.-Y.; Dong, B.; Guo, B.-Y.; Liu, D.-P.; Liu, C.-G.; Chai, Y.-M. Hydrogen evolution under large-current-density based on fluorine-doped cobalt-iron phosphides. *Chem. Eng. J.* **2020**, *399*, 125831. [CrossRef]
73. Huang, C.; Zhou, J.; Duan, D.; Zhou, Q.; Wang, J.; Peng, B.; Yu, L.; Yu, Y. Roles of heteroatoms in electrocatalysts for alkaline water splitting: A review focusing on the reaction mechanism. *Chin. J. Catal.* **2022**, *43*, 2091–2110. [CrossRef]

MDPI AG
Grosspeteranlage 5
4052 Basel
Switzerland
Tel.: +41 61 683 77 34

Coatings Editorial Office
E-mail: coatings@mdpi.com
www.mdpi.com/journal/coatings